CONTENTS

INTRODUCTION

The essential role of the nervous system is to coordinate the body systems by means of electrical impulses. It is an extremely efficient system, not only in terms of its speed of operation, but also in terms of its low energy and nutrient consumption.

It is the **organisation of the nervous system** that results in the efficient coordination of the body as a whole. The nervous system is divided into the **central and peripheral systems.**

The central nervous system **(CNS)** is analagous to a telephone exchange. It receives an electrical input from all over the body, and It outputs electrical information to all parts of the body. The peripheral nervous system **(PNS)** is analagous to the wires connecting the telephone exchange with the individual telephones. Thus it defines the routes that the electrical impulses follow, linking all the organs of the body with the central nervous system.

It is the **specialisation of the neurones** (nerve cells) that results in the efficient transmission of electrical impulses, and therefore provides the answer to how the system works. The neurones are linked together like wires in an electrical system. There are three types of circuit, or neural pathways.

1. Link Circuits

These are the neurones that link the input neurones (**sensory pathways**) and output neurones (**motor pathways)** of the **PNS** to specific parts of the **CNS**. Impulses flow rapidly through specific circuits. The circuit would be broken if one link was damaged. This happens in paraplegics, where damage to the spinal cord (part of the CNS) breaks the link to the motor pathways in the limbs.

2. Local Circuits

These are short neurones found in the **CNS** whose major connections are made with the immediate environment. There will be many such local circuits in the brain (the major part of the CNS) used in such activities as problem solving and idle day dreaming, as well as many brain activites that we are not aware of.

3. Single Source/Divergent Circuits

Here one neurone in the **CNS** diverges to have a large number of connections into many link circuits, thus integrating the many activities of the nervous system. These circuits would be used when brain activity in local circuits results in a response via the **PNS**, as would occur in solving an examination question and writing the response.

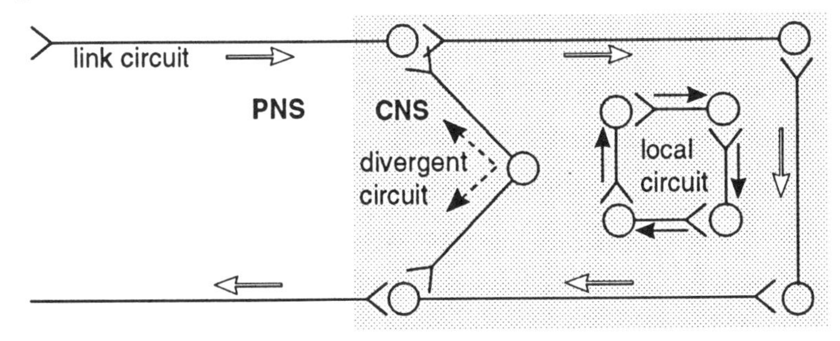

DIAGRAM OF THE NERVOUS SYSTEM

- *Only those parts that are mentioned in the unit have been labelled.*

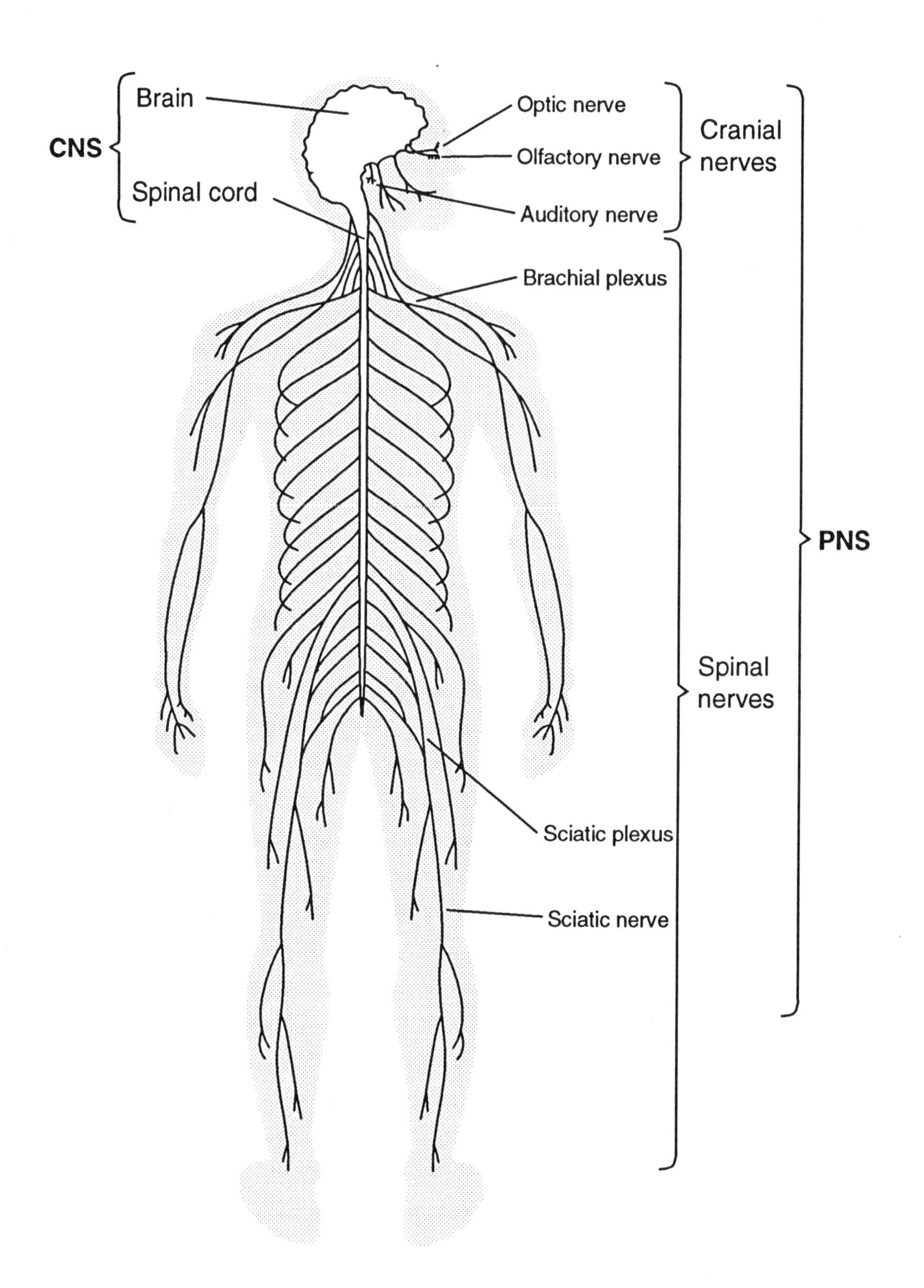

NERVOUS TISSUE

Nervous tissue is divided into three types of cells:

1. **Neurones.** These are the cells that conduct the electrical impulses.

2. **Neuroglial Cells.** These are specialised connective tissue cells that either support, protect, or improve the efficiency of the neurones in the CNS. **Astrocytes** are star shaped neurogleal cells that form sheaths arround the capillaries surrounding the brain. They form a **blood brain barrier**, allowing rapid diffusion of small molecules like oxygen, carbon dioxide and alcohol. Larger molecules diffuse slowly or not at all.

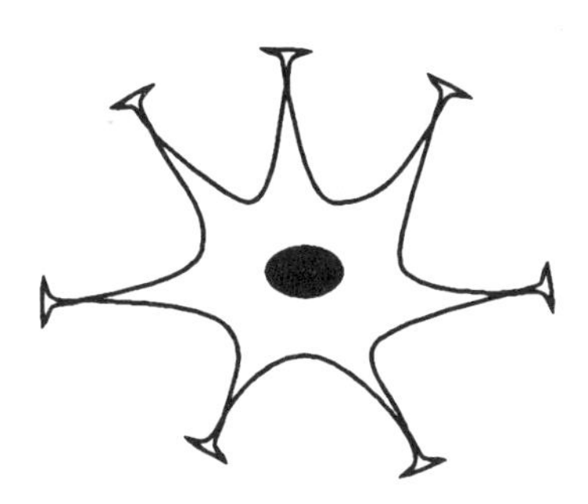

3. **Schwann Cells.** These are connective tissue cells that support, insulate and speed up the conduction of impulses in the neurones in the PNS. They are thin flat cells that wrap arround the fibres of the neurones, resulting in the insulating **myelin sheath** that consists of several layers of plasmamembrane.

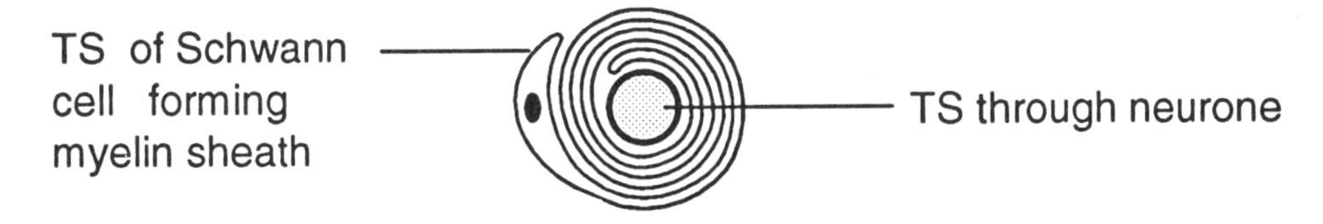

Neurones.

Neurones are nerve cells that are specialised for rapid co-ordination by transmitting electrical impulses. Structural specialisation in neurones modifies the typical cell in a number of ways.

A typical cell has a nucleus and cytoplasm limited by a plasma membrane. The cytoplasm contains various organelles such as mitochondria, granular endoplasmic reticulum, and filaments.

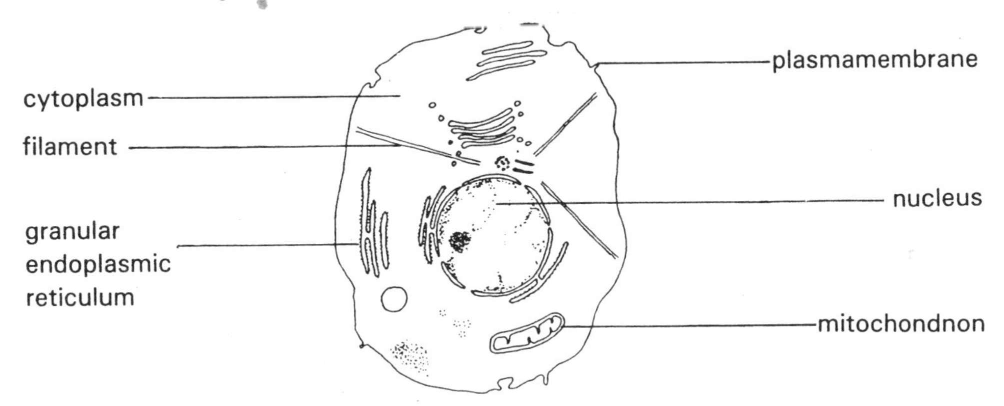

- *Check that you know the functions of these organelles.*

NERVOUS TISSUE (Cont.)

Structure of the Neurone

The modifications from the basic cell type in neurones are:

1. The cytoplasm is organised into long processes, known as **nerve fibres,** to increase the distance over which the impulse is conducted. This may increase the length of a single cell to as much as a metre! Fibres contain filaments and fine tubules called **neurofibrils.**

2. The nucleus and surrounding cytoplasm is the **cell body.** It contains a lot of granular endoplasmic reticulum seen in the light microscope as dark granules called **Nissl granules.**

3. The nerve fibres are of two types:

Axons carry impulses away from the cell body. There is only one axon per neurone, but the ends branch into many small nerve endings.

Dendrons carry impulses towards the cell body. There may be one or more dendrons (or dendrites if they are small) per neurone.

4. The nerve endings of the axon end in swollen areas, characterised by the presence of mitochondria and secretory vacuoles. These are the **synaptic knobs**, or the **motor end plates** (if the axon ends in a muscle).

5. Impulses are initiated in receptor regions of the dendron, or dendrites. The receptor regions are characterised by specific proteins in the plasmamembrane known in neurones as the **neurolemma.**

Structure of a Motor Neurone

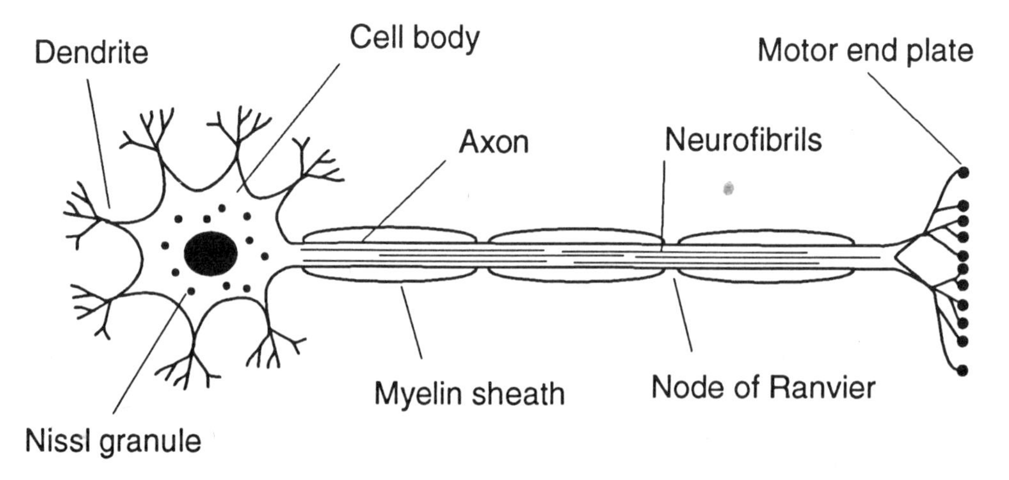

NERVOUS TISSUE (Cont.)

Classification of neurones.

Neurones are classified according to their srtucture or their function.

A. Structural Types. This depends on the position of the cell body in relation to the fibres.

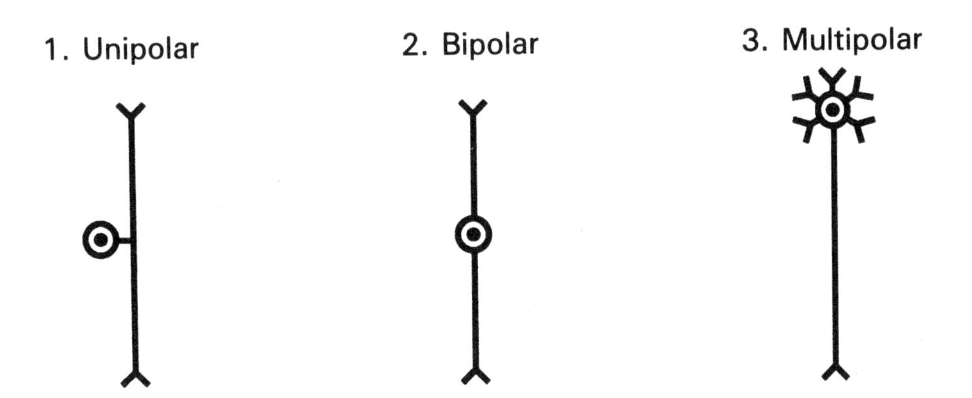

B. Functional Types. Neurones are more commonly classified according to the path of the impulse in relation to the CNS:

Sensory or afferent neurones conduct impulses towards the CNS.

Motor or efferent neurones conduct impulses away from the CNS.

Interneurones or connecting neurones connect the afferent and efferent pathways

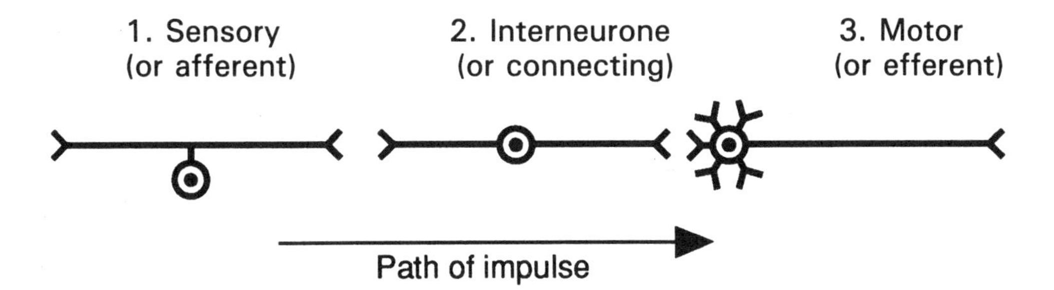

The relationship between adjacent neurones is important in the organisation of the nervous system. There may be many interneurones between the sensory and motor neurones, or none.

PHYSIOLOGY OF NERVE CELLS

Physiological specialisation is the way the neurones behave compared to typical cells, and results in the conduction of the electrical impulse. To understand the way the neurone works it is helpful to review some features of typical cells.

The internal environment of the cell is separated from the external environment by the plasma membrane. Since the membrane is partially permeable there will be some differences between these environments.

Inside the cell there will be large negatively charged protein ions, too large to pass out of the cell. There will also be small positively charged potassium ions. These would diffuse out of the cell but for the attraction of the protein ions, since unlike charges attract.

Outside the cell in the tissue fluid there are also ions, derived mainly from salt, sodium chloride. The positively charged sodium ions diffuse slowly into the cell, but are pumped out again by active transport. The sodium pump uses energy generated by the cell. The negatively charged chloride ions would diffuse into the cell but for the attraction to the sodium.

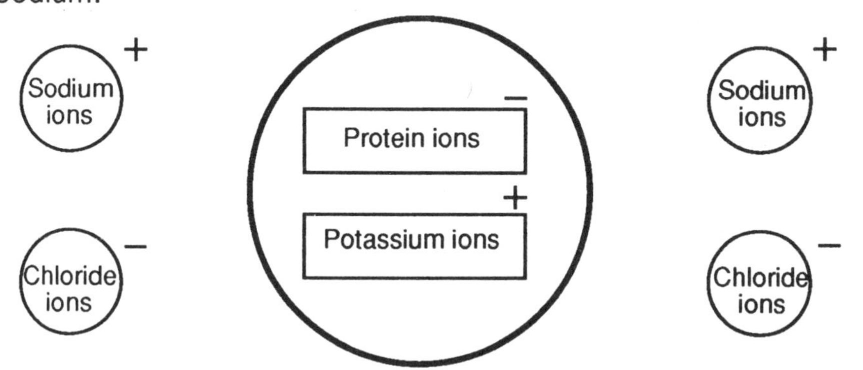

The balance between the attraction of unlike charges, diffusion and active transport results in more sodium than chloride on the outside of the cell, and more protein that potassium on the inside of the cell. This means there is a net positive charge on the outside of the membrane, and a net negative charge on the inside.

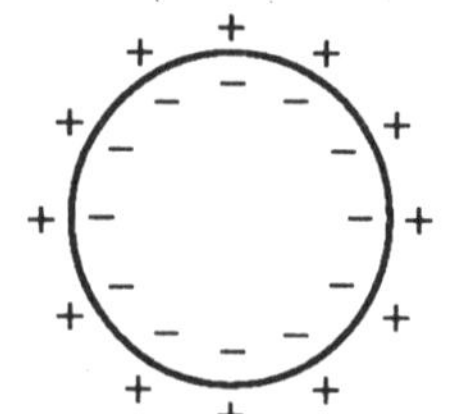

The area of positive ions is separated from the area of negative ions by the plasma membrane. The membrane is said to be **polarised.**

If the membrane were to be removed negative and positive ions would move together, i.e. they would **perform work**. With the membrane in place there is an **electrical potential for doing work.** This potential exists in all cells. It is called the **resting potential** and is measured in volts, or millivolts because it is a small amount of electrical work. All cells have a resting potential, but it is somewhat higher in neurones. The resting potential in neurones is about **-70mV**.

THE NERVE IMPULSE

The physiological specialisation of neurones that changes the resting potential into an impulse is due to the fact that the membrane can change its permeability to sodium when suitably stimulated.

This is brought about by membrane proteins that form sodium ion channels. These are normally closed, but they open by changing their shape in response to specific environmental changes.

The effect of sodium ions rushing into the neurone changes the balance between negative and positive ions, and the potential inside the cell changes from **-70mV** to **+40mV**. The membrane is said to be **depolarised,** and the resting potential has changed to an **action potential.**

The resting potential is quickly restored by potassium ions rushing out, through potassium ion channels, and more slowly by the pumping out of sodium. The membrane is then **repolarised.**

- *The diagrams below are intended to represent the changes in net charges in a section across a nerve fibre when the membrane is stimulated to change its permeability. Complete the second two, using the diagram on the next page to help if necessary.*

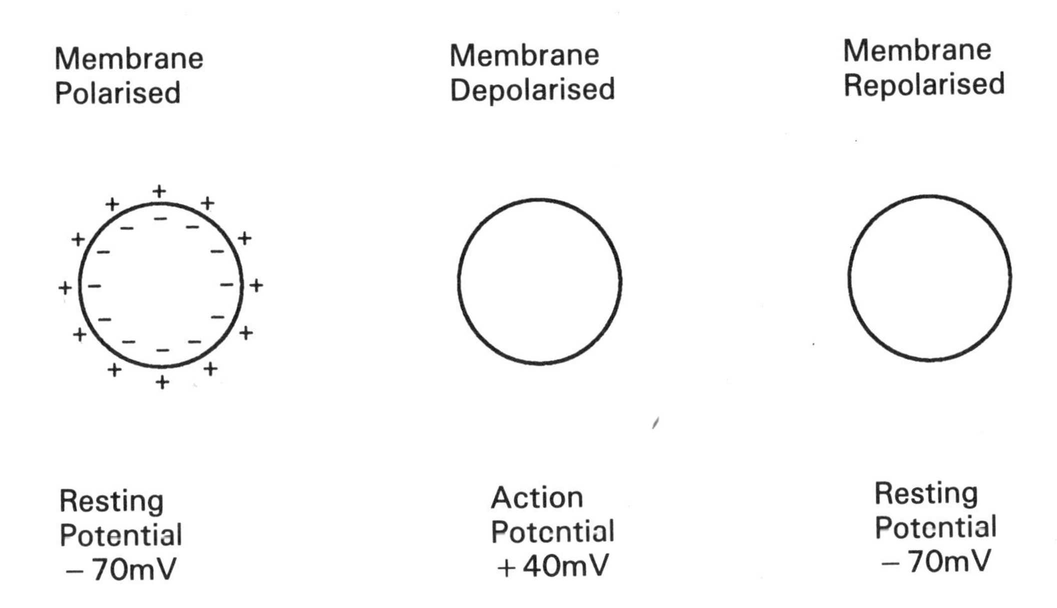

The action potential only lasts a short time - about one **millisecond.** However the effect of the action potential is to change the sodium permeability of the adjacent membrane, setting up another action potential. When the membrane at one end of a neuron is stimulated to produce an action potential it sets off a propagation of action potentials along the entire length of the neurone.

This then is the impulse - a sequence of self-propagating action potentials.

THE NERVE IMPULSE (Cont.)

Graph to show the change in voltage with time during one action potential.

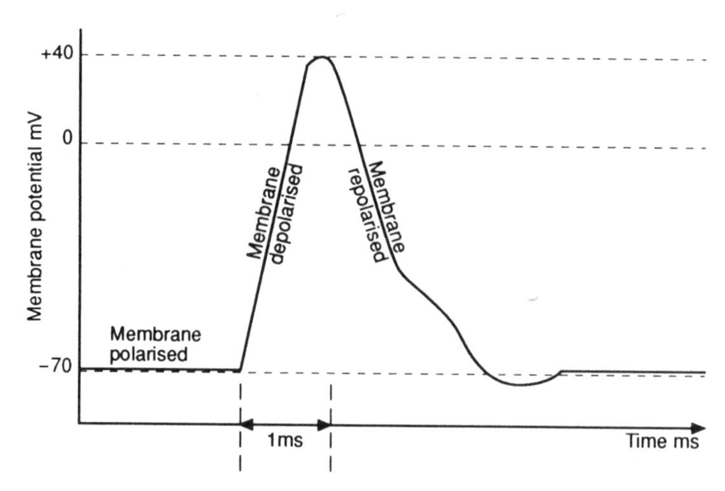

Section along a Nerve Fibre to show the changes associated with the Action Potential.

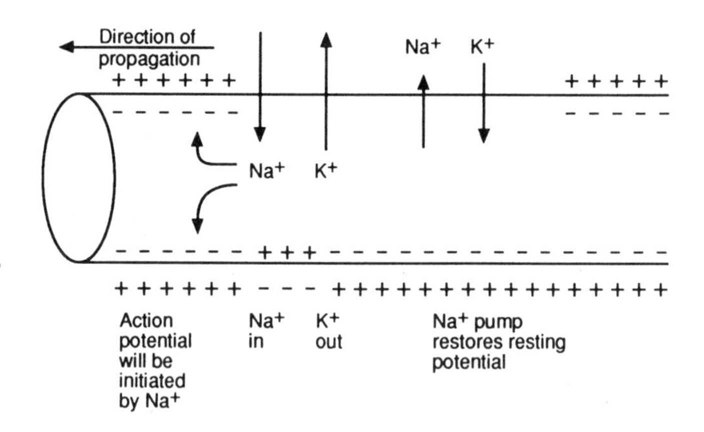

Detail of the changes in the Plasma Membrane associated with the Action Potential.

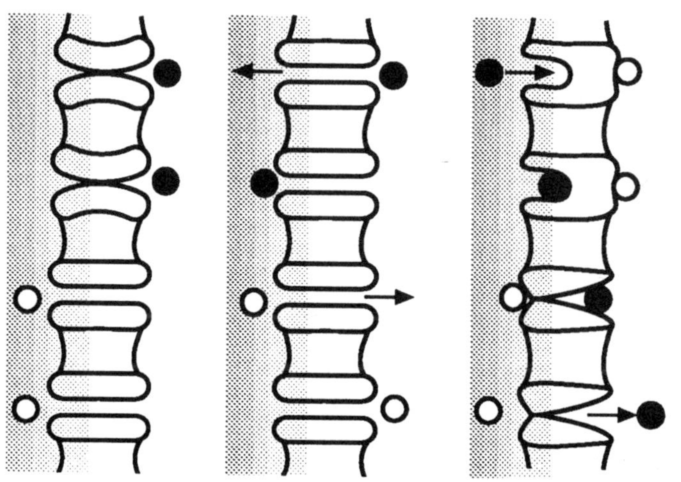

● *Label this diagram from the information above and on page 9*

THE NERVE IMPULSE (Cont.)

The potential differences in the neurone can now be clearly defined.

The resting potential (-70mV) is the voltage difference between the inside and the outside of the cell in the absence of stimulation.

The action potential (+40mV) is the voltage difference between the inside and outside of the cell when stimulated sufficiently to produce a change in the membrane permeability to sodium.

The effectiveness of the action potential for co-ordination in the nervous system depends on a variety of factors that need a closer examination.

The All or Nothing Law.

A small stimulus will not initiate an action potential. When the stimulus reaches a certain level then an action potential is fired. This is the **threshold** level. At this level of stimulation an action potential of +40mV will be fired. If the level of stimulation is raised beyond the threshold level the action potential that is fired will still only be +40mV.

The Refractory Period.

Once the membrane has been depolarised there is a period of time during which it cannot be depolarised again. Therefore a second action potential cannot be fired and the membrane is said to be unexcitable. This is the refractory period. The time varies with the stimulus strength. For a stimulus at the threshold level this is about 5 milliseconds. At a higher stimulus the refractory period can be reduced to 1 millisecond.

Rate of Firing (Frequency)

It can be seen from the refractory period that the rate of firing will vary with stimulus strength.

Strong stimulus - fast rate, (high frequency)
Threshold stimulus - slow rate, (low frequency)
Below threshold stimulus - no action potentials fired

Speed of Conduction.

The action potential depends upon the change in membrane permeability to the flow of sodium ions which then produces the net positive change in the cytoplasm. So it is the **resistance** of the membrane and the **volume** of the cytoplasm that produce a time factor, and speed of conduction will relate to the surface area to volume ratio in the neurone.

Small amount of cytoplasm in relation to the membrane, resistance high, conduction slow.

Large amount of cytoplasm in relation to the membrane, resistance low, conduction fast.

The myelin sheath increases the speed of conduction because the action potentials hop from node to node. This is known as **saltatory conduction.**

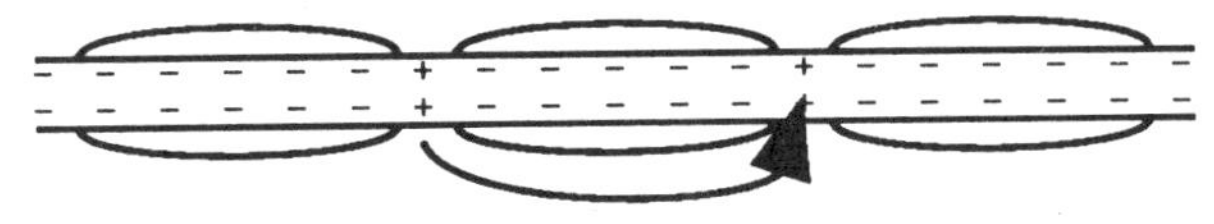

THE INITIATION OF ACTION POTENTIALS

The action potentials are not normally initiated along the fibres. They are initiated in the **receptors.** These are areas where the resting potential can be changed gradually. A small stimulus results in sodium ions moving into the cell gradually increasing the potential difference. This is known as the **generator potential**, and it is non-conductive, i.e. it does not change the properties of the adjacent membrane. When the generator potential reaches a certain level it fires an action potential which is then conducted.

Diagram to show the initiation of action potentials in a sensory neurone.

generator region

conductive region

NA+

stimulus – may be mechanical, electrical, or chemical.

Increased permeability to sodium ions increases generator potential.

When the threshold level is reached the action potential is fired.

Action potential is propagated along the fibre.

If the generator potential is maintained a second action potential will be fired, after the refractory period.

stimulus	generator potential	action potentials
small	threshold level reached slowly	low frequency
large	threshold level reached quickly	high frequency

Adaptation.

If a stimulus is maintained over a period of time, the generator potential gradually declines, and action potentials cease to be fired. This serves to protect us from being aware of a continuous but unpleasant stimulus, such as the tickly shirt!

THE ACTION POTENTIAL AT THE NERVE ENDINGS

The action potential initiated in the receptor will be propagated along the axon of the neurone, and along each of the terminal branches ending in a synaptic knob.

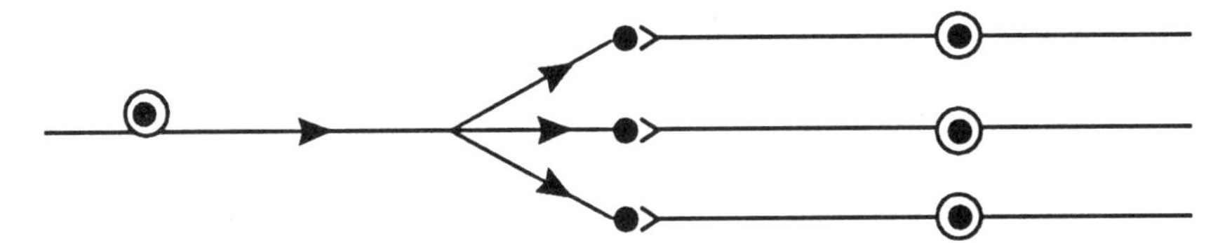

There will be a small gap between the synaptic knob and the adjacent neurone, known as the **synapse.** Since action potentials can only be propagated along a continuous membrane they cannot cross the synapse. However, the action potentials produce changes in the knob which may result in a new action potential being initiated.

The Structure of the Synapse.

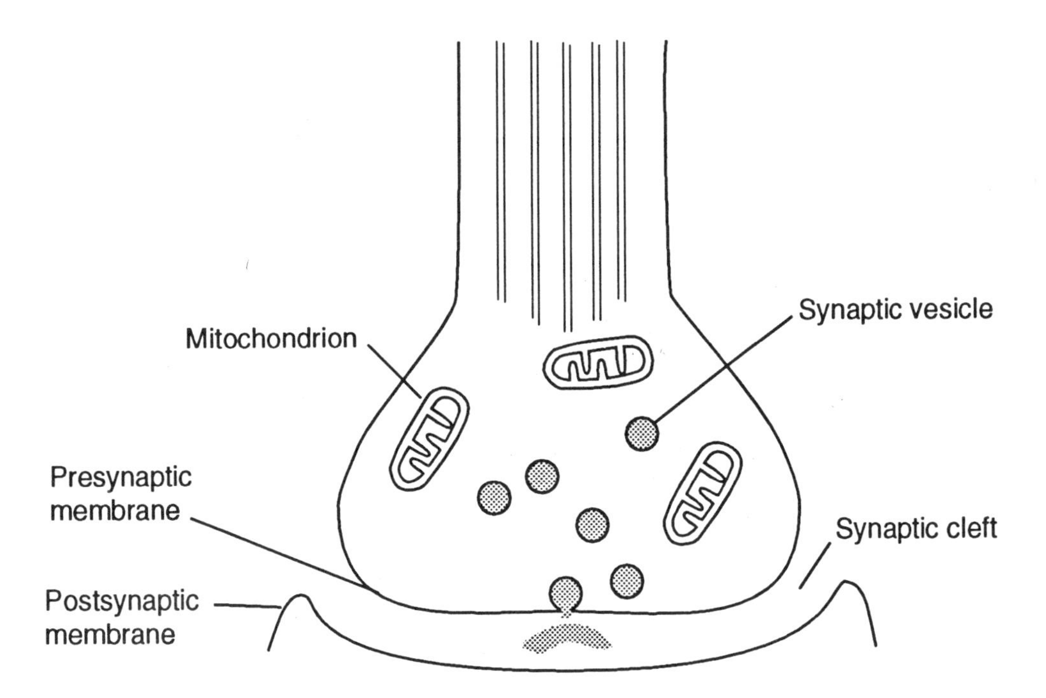

The effect of the action potential on the synaptic knob is:

1. It increases the permeability of the pre-synaptic membrane to **calcium ions**

2. The calcium ions cause some of the **vesicles to fuse** with the presynaptic membrane.

3. The vesicles **release the transmitter substance** into the synaptic cleft.

THE ACTION POTENTIAL AT THE NERVE ENDINGS (Cont.)

The Effect of the Transmitter Substance

1. The transmitter substance activates receptor sites on the post synaptic membrane. The receptors are specific plasma proteins.

2. Either there is a gradual increase in permeability to sodium ions producing a change in the membrane potential of the post synaptic membrane. This is known as the **excitory post-synaptic potential (EPSP)**. When the threshold level is reached an action potential is fired.

Or there is an increase in permeability to potassium ions which pass out of the post-synaptic neuron, increasing the negativity of the membrane potential. No action potential occurs. This change in the membrane potential is known as the **inhibitory post-synaptic potential (IPSP.)**

3. The effect of the transmitter substance is short lived because an enzyme in the synaptic cleft breaks it down, and the products are transported back into the synaptic knob for resynthesis, or it is transported back into the synaptic knob directly.

Recycling of transmitter substance.

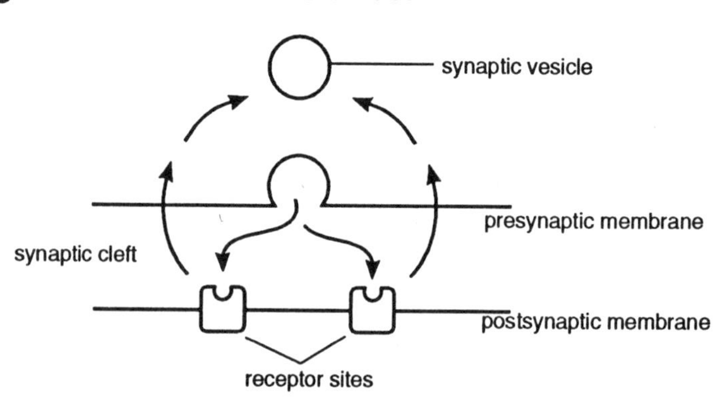

Whether a synapse is excitory or inhibitory is a function of the receptor site. Any one synapse is therefore always either inhibitory or excitory.

- *Note that the receptor region of the inter and motor neurones will be the post-synaptic membranes.*

Summation

One post-synaptic neurone will have many synaptic connections with a number of different pre-synaptic neurones. An action potential will only be fired if the summative effect of all the synapses reaches a threshold value. Inhibitory synapses can cancel the effect of excitory synapses. Thus conflicting or small inputs will be filtered out at the synapse.

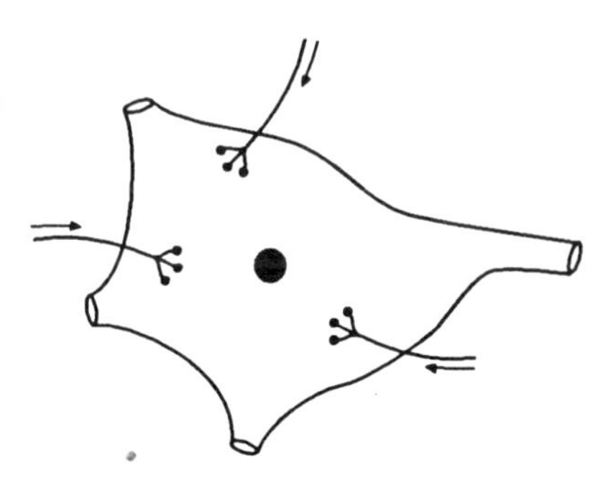

THE ACTION POTENTIAL AT THE NERVE ENDINGS (Cont.)

Nerve Ending of Motor Neurones

Motor neurones end at effector organs, either glands or muscles. Instead of a synaptic knob there is a motor end plate, but the structure is similar.

Structure of the Motor End Plate

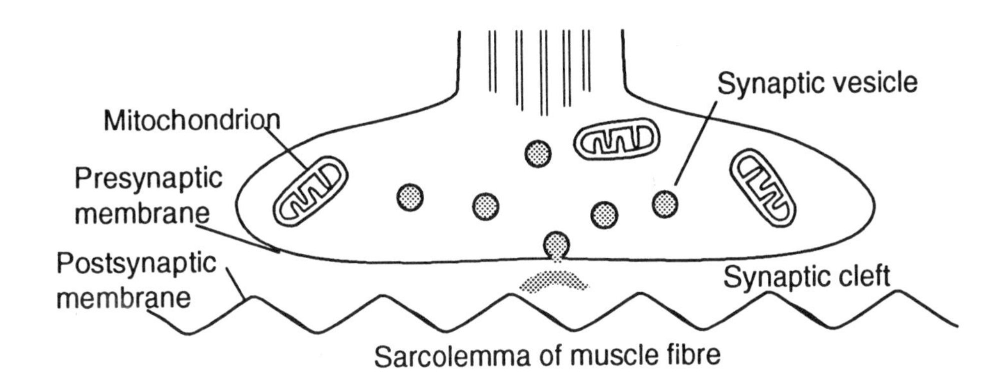

Muscles are capable of being depolarised like neurones, so the system operates in a similar way to the synapse:

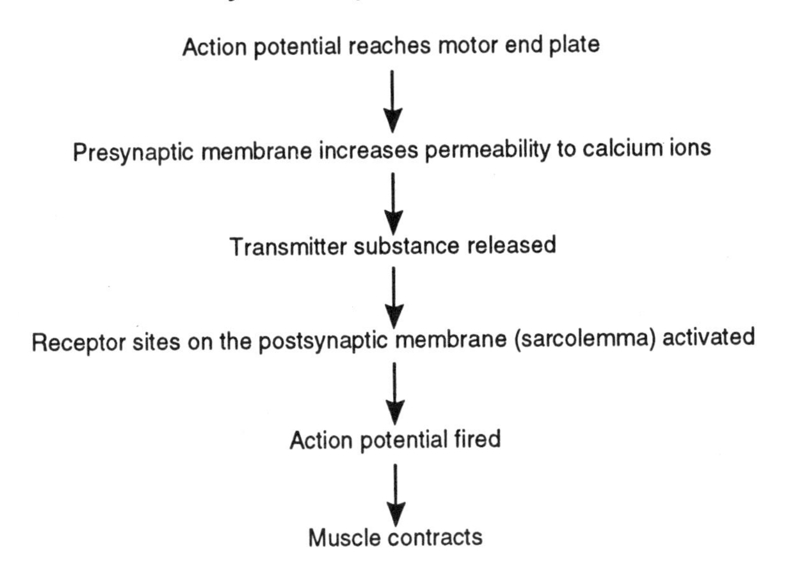

Common Transmitter substances

Acetyl choline - found at motor end plates of muscle - broken down by the enzyme choline esterase.

Noradrenaline - found at end plates in smooth and cardiac muscle.

- *There are many other transmitter substances , particularly in the brain, but it is worth noting that the effect of all transmitter substances is to be excitory, or inhibitory. In other words the transmitter substance acts as the stimulus that initiates action potentials or prevents the action potential from being fired.*

ADAPTATION AT THE SYNAPSE

Temporary adaptation.

If the frequency of the action potentials at the synapse is high, or maintained over a long period then the transmitter substance may be secreted faster than it can be recycled. This would lead to a temporary failure to fire action potentials in the post-synaptic neuron - this could be another reason for not noticing that tickly shirt!

Regulatory adaptation.

If a synapse is subject to long term changes in the amount of transmitter substance released, induced by disease or drugs, normal synaptic transmission may be maintained by modification of the post-synaptic membrane. This is shown in the diagram below.

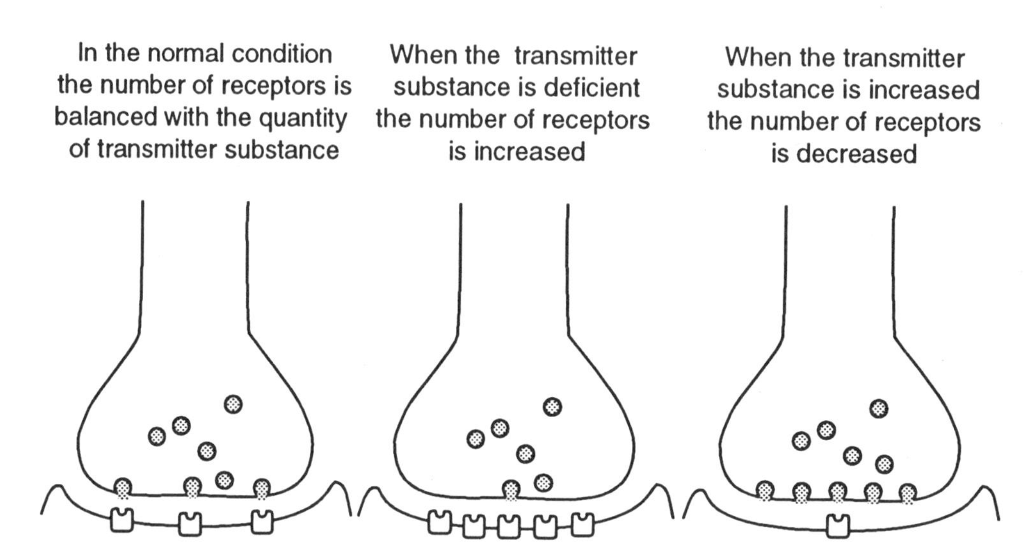

The ability of the neurone to modify the synapse chemically and physiologically in response to change in use or environment, is a key factor in adapting the nervous system to meeting the needs of the body.

Drugs can affect the synapse in different ways. **Curare,** the poison used in the arrow heads of the South American Indians, interferes with the ability of the transmitter substance depolarising the post-synaptic membrane. **Strychnine** increases the effect of the transmitter substance by interfering with the enzyme that normally breaks it down.

In addition to modifying existing synapses new synapses can develop between neurones to establish new circuits. This way specific responses can be produced more quickly to commonly occuring situations. Neurones in the young are more modifyable than adults.

Fetal neurones have been successfully transplanted into the brains of patients suffering from Parkinson's disease. The donor fetal neurones establish synaptic connections with the recipient's cells.

THE CENTRAL NERVOUS SYSTEM

Gross Structure

It is a common phenomenon in moving animals to locate major sensory receptors at the front end, i.e. the end going first into the new environment. Thus the CNS develops most at this end to receive and process the information. This is the **brain**. The less complex part of the CNS is the **spinal cord.**

The brain is further divided into a **fore - mid - and hind-brain,**receiving the input from the three major sense organs, the nose, the eyes, and the ears. This is the order that the three major senses appear in, for instance, the dog.

Mostly when we think of the brain we equate it with intelligence. This function is attributable only to the roof of the forebrain, which has developed to the extent that it takes up most of the space in the cranium.

Development of the CNS

The CNS develops from a tube of nervous tissue. Blood vessels supply nutrients to the outside of the tube, and fluid within the tube (cerebro-spinal fluid) helps the circulation of nutrients and the removal of waste products. Although some areas develop more than others the essential tubular arrangement is not lost.

The diagrams below show the basic pattern of development found in all animals with a backbone, such as the dog.

1. CNS as a tube of nervous tissue.

2. Specialisation into brain and spinal cord.

3. Brain develops into fore mid and hind brains.

4. Development of functional areas.

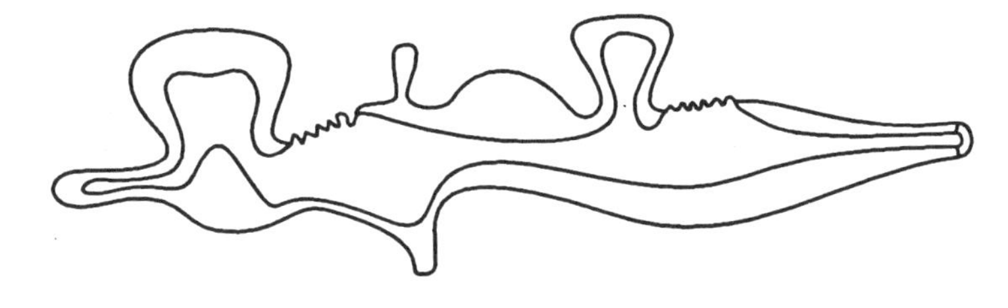

DEVELOPMENT OF THE HUMAN BRAIN

The CNS develops in the embryo from a group of cells which invaginate from the outer layer to form a tube. These cells can be picked out as a dark streak in embryos, and is referred to as the primitive streak. Test tube embryos are not kept viable beyond this stage.

Once the tube has formed the brain differentiates into the three bulges which will give rise to the fore - mid - and hind-brains. The brain flexes in the region of the midbrain to accommodate the upright stance.

By the time the baby is born the roof of the front part of the forebrain has developed into the two cerebral hemispheres which envelop the rest of the brain.

- *Use four colours to distinguish the three areas of the brain and the spinal cord in the diagrams below:*

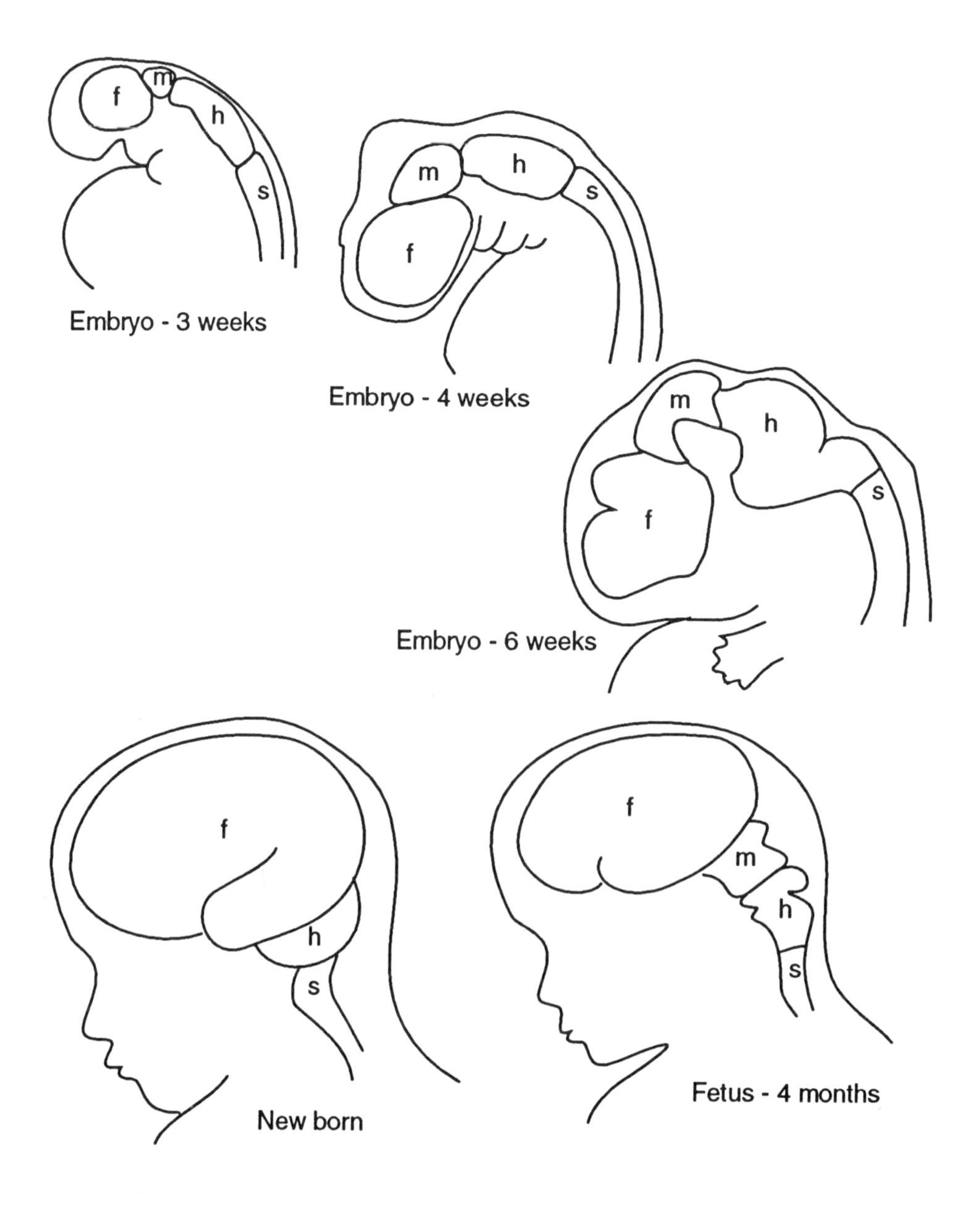

SECTION THROUGH THE BRAIN AND SPINAL CORD

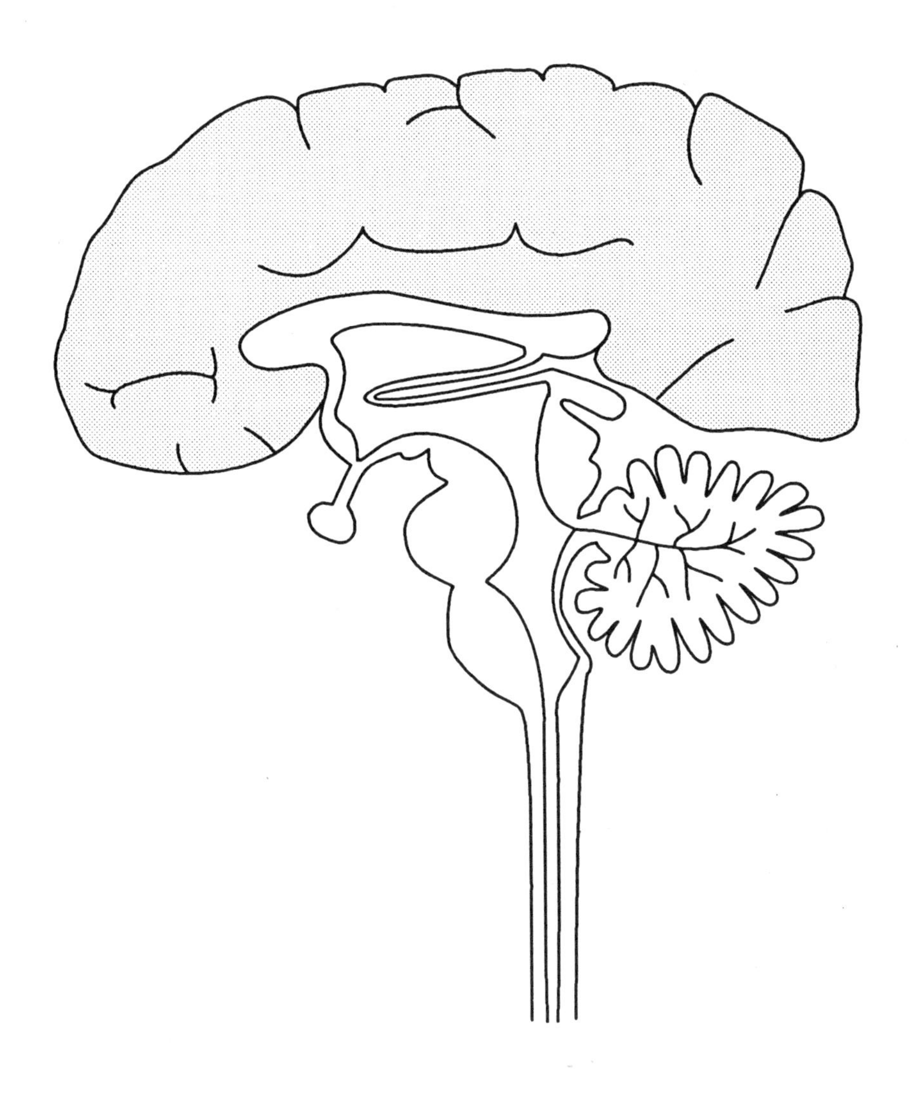

- *Distinguish the fore- mid- and hind-brains in this sectional view, using the same colour code as in the previous page.*

- *Label the parts of the brain from a text book, making sure thatyou have labelled all the parts referred to in the unit. Notice that the shaded area represents the outer surface of the right hemisphere. This would not be cut through in this section.*

- *Colour the cerebrospinal fluid in blue. Outline the the whole of the central nervous system in red, to indicate the blood supply. Notice the anterior and posterior choroid plexus where the blood comes close to the cerebrospinal fluid.*

PROTECTION OF THE BRAIN

The brain is protected from severe physical damage by the bones of the cranium. However the bones of the cranium are hard, and could damage the nervous tissue when the head is moved. A fluid filled space between the cranium and the brain acts as a shock absorber during normal movement. It is not, however, adequate to absorb the shock of blows to the head that a boxer might experience.

The fluid is found between the meninges, membranes that also carry blood vessels. Inflammation of these causes meningitis.

Section to show structures protecting the brain.

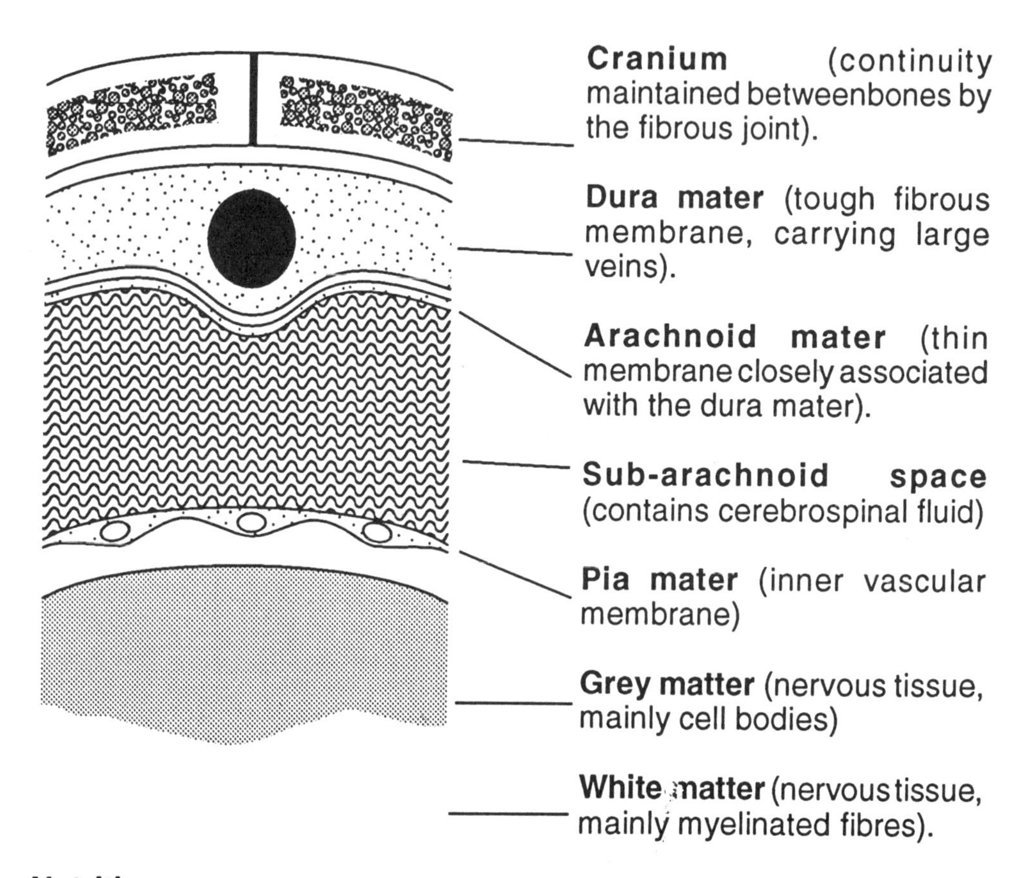

Nutrition

Nutrients are supplied to the brain by the blood vessels of the pia mater, with selective transport being maintained by the **blood brain barrier.**

The **choroid plexus** is where the pia mater comes into contact with the neural lobe, so it is here that the blood supply produces the cerebrospinal fluid of the ventricles (neural tube of the brain) and the spinal canal (neural tube of the spinal cord).

Fluid flow is maintained by outlets in the hind brain and the base of the spine into the sub-arachnoid space.

From the arachnoid space the fluid carrying waste flows into the veins of the dura mater.

THE CEREBRAL HEMISPHERES

The surface of the cerebrum is folded. This increases the capacity of this area of the brain. The biggest of these folds divides the cerebrum into the right and left **cerebral hemispheres.** Each hemisphere is further divided into four other major lobes.

Diagram of the Left Hemisphere.

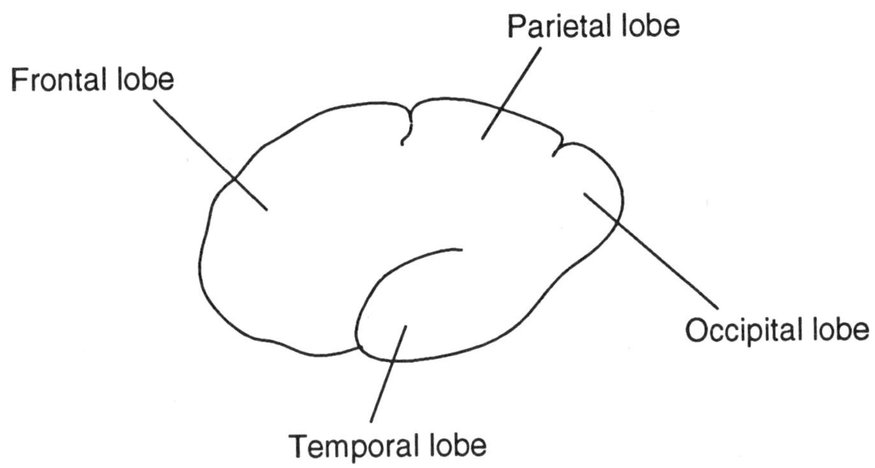

- *This is a useful diagram to learn to draw. Note that it looks a bit like a boxing glove with the thumb pointing to the front.*

Cerebral cortex

This is the outer layer containing the unmyelinated neurones of the grey matter. This is where there are many interlinking interneurones associated with intelligence.

Some functional areas have been mapped out from observations on brain damaged people.

frontal lobe controls conscious movement.

parietal lobe interprets sensory information from the skin and proprioceptors of the skeletal muscles

occipital lobe interprets the sensory information from the eyes

temporal lobe interprets sensory information from the ears.

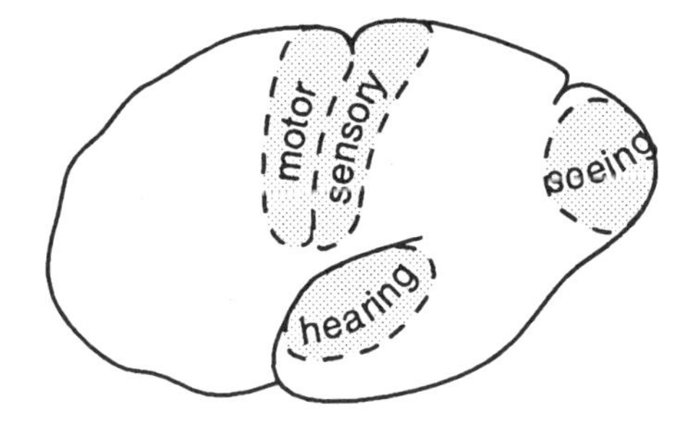

Between these areas are the association areas where thinking, reasoning, decision making, memory and emotion happen, as a consequence of linking the various centres.

THE CEREBRAL HEMISPHERES (Cont.)

The Role of the White Matter

The myelinated fibres of the white matter provide the links between the major areas of the cortex. There are three types:

1. Fibres that link areas within one hemisphere.
2. Fibres that link the two hemispheres. This forms the **corpus callosum.**
3. Fibres link the cortex with the other regions of the brain and spinal cord.

Specialisation between the hemispheres

Diversity of function in the cerebrum is further increased by specialisations in the left and right hemispheres. The left is better for literacy skills and sequential processing. The right is better for spatial relationship skills and creativity.

Speech is controlled by two areas in the left hemisphere.

Wernicke's area is important for comprehending language. It also initiates the choice of sounds for producing meaningful speech.

Broca's area is linked to Wernickes area, and is important for co-ordinating the muscular responses that produce speech.

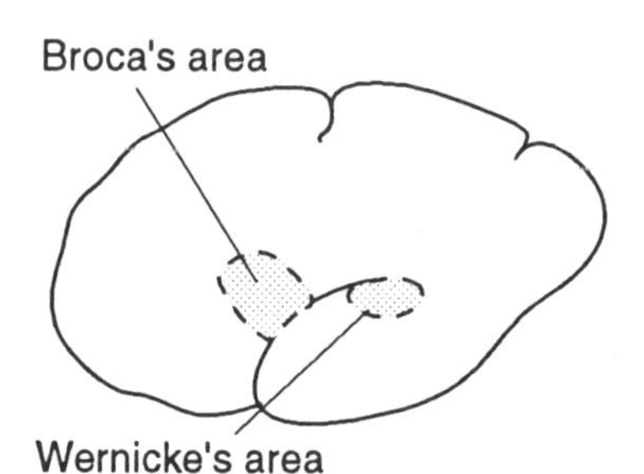

Dominance

Because fibres cross between right and left at specific places in the CNS functions of one side of the cortex relate to body structures on the other. Thus it is left side of the cortex that controls muscular activity in the right hand, and vice versa.

In right handed people the left hemisphere dominates the right hemisphere, and skills controlled by the left hemisphere are likely to be superior.

In left handed people the right hemisphere dominates the left hemisphere, and skills controlled by the right hemisphere are likely to be superior.

Braille represents letters by using raised dots on a grid of six. Interpretation of the letter requires the reader to feel the dots and assess the spacial relationship between them. Many blind people prefer to use the left hand, because the sensory neurones link with the cortex of the right hemisphere. This is irrespective of whether they are left or right handed.

- *You could try this out on blind folded subjects making your own braille from pins stuck into cork. To make the task easier you could use a grid of four.*

THE CEREBRAL HEMISPHERES (Cont.)

The sensory area of the parietal lobe.

Mapping of this sensory area shows that it can be functionally divided into the areas of the body from which the sensory information comes, and that this mapping reflects the relative sensitivity of the different parts of the body.

Functional zones of the Sensory Area

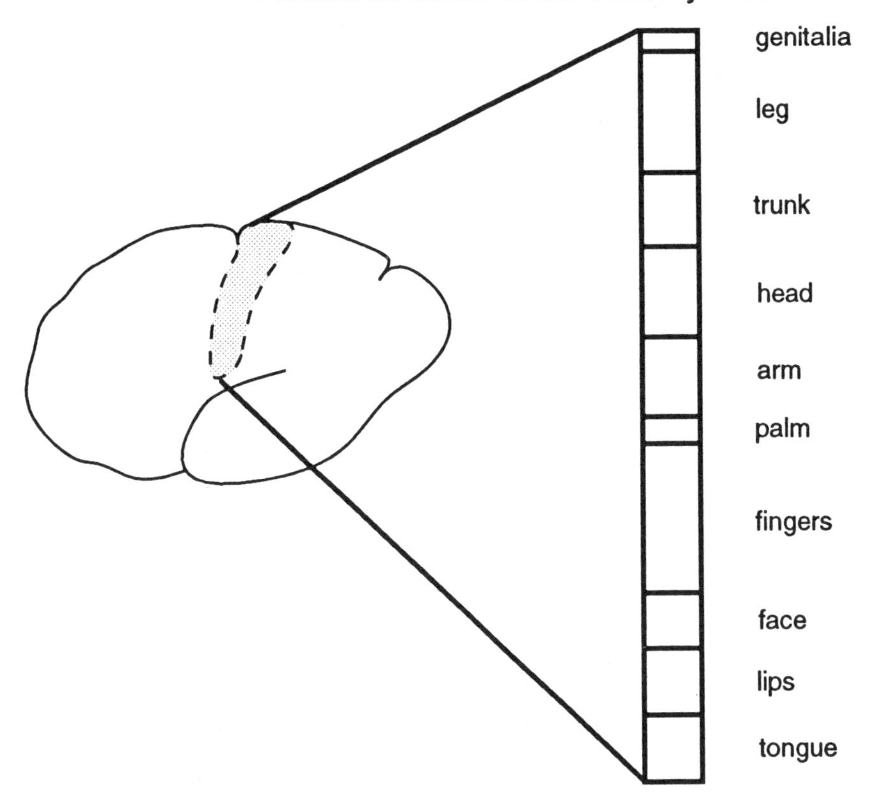

Specific neural pathways carry impulses initiated by the receptors to the sensory cortex. Since the pathways cross over at some point, then impulses initiated on the right side of the body are interpreted on the left side of the brain.

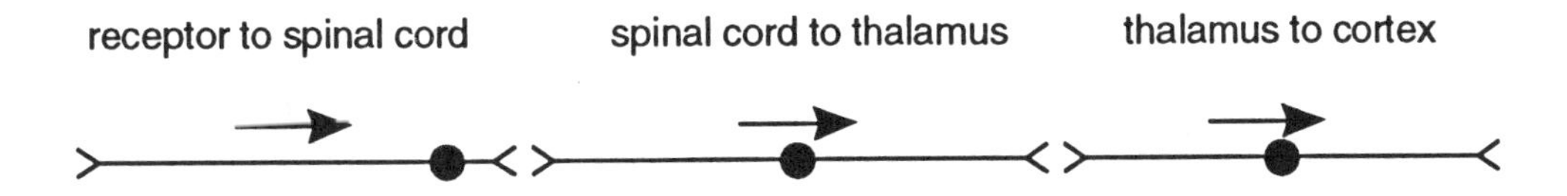

Note that the sensory cortex interprets, but the neural path way locates the sensory area. Thus if a leg is amputated, and the cut end of the neural path way at the stump that originally connected to the big toe is stimulated sufficiently to initiate an impulse, this will be interpreted in the cortex as coming from the big toe. A curious sensation if you know you have no big toe!

THE CEREBRAL HEMISPHERES (Cont.)

Motor Area of the frontal lobe

The motor area of the cortex can be mapped in a similar way to the sensory cortex. Action potentials will be initiated as a consequence of brain activity in other areas of the CNS, which have a variety of inhibitory and excitory synaptic connections with the motor cortex.

The neural pathway linking the motor cortex to the skeletal muscles is via splendidly long neurones, **the pyramidal cells.** Fibres from the pyramidal cells produce conspicuous descending pathways down the spinal cord. Most fibres cross over at some point, so that muscular activity on the **right** side is controlled from the **left** hemisphere.

There are thus only 2 neurones linking the cortex to the muscle, with one synapse and one motor end plate.

- *What effect will this have on the speed of transmission?*

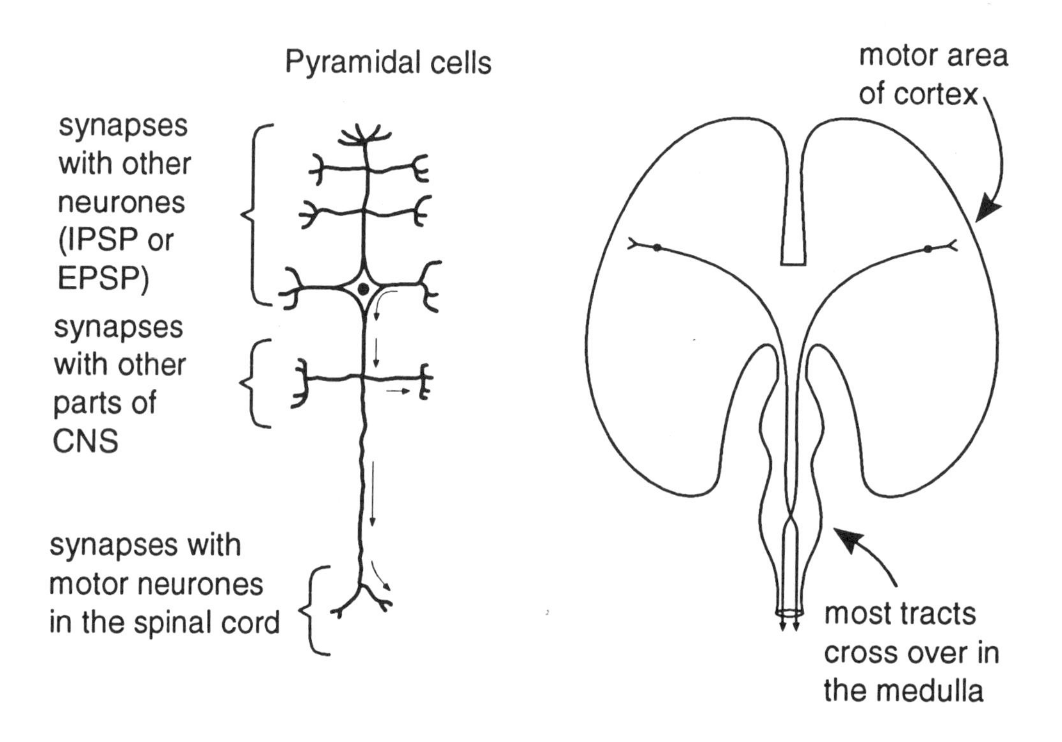

The **extra pyramidal tracts** synapse with the **thalamus** and the **cerebellum,** and run along the pyramidal tracts in the spinal cord. These extra pyramidal tracts provide the neural pathways that are involved with the many muscular activities that require modification following a simple decision to, for instance, kick a ball.

- *Try kicking a ball without using any muscle except those in the kicking leg.*

There are also **inhibitory tracts** that can override spinal reflex responses.

FUNCTIONS OF THE BRAIN

Cerebral Hemispheres

This region results in:-

1. Conscious thought.
2. Higher levels of intelligence associated with human behaviour, such as reasoning, judgment, and will power.
3. Interpretation of senses.
4. Initiation of muscular responses.
5. Emotional response that assesses such values as beauty, virtue and sin.
6. Domination of reflex actions, such as coughing.

Thalamus

Located laterally just above the hypothalamus, this part of the forebrain acts as a relay station. There are an enormous number of synapses linking the cerebral circuits with other parts of the CNS.
The main links are:-

1. Sensory input to cerebrum from the eye, the ear and the skin.
2. Motor output from motor areas of cerebrum to the cerebellum the centre of co-ordination of muscular responses.

Hypothalamus

The hypothalamus is vital to **homeostasis.** by means of the following functions:-

1. It contains sensory receptors to detect the osmotic presure and the temperature of blood and initiates the reponse that maintains a constant osmotic pressure of the blood, and body temperature.
2. It controls the anterior lobe of the pituitary by means of a capillary network that carries substances from the hypothalamus to stimulate hormone release.
3. The posterior lobe of the pituitary gland is actually an extension of the hypothalamus, so its hormone secretion is under direct nervous control.
4. It controls those peripheral motor nerves that are not under conscious control. These are the nerves of the autonomic nervous system that control respiration, circulation and digestion, instinctive behaviour patterns and temperature regulation.

FUNCTIONS OF THE BRAIN (Cont.)

Mid Brain.

1. The roof of the mid brain is swollen into the optic lobes. These have some control over eye movement, but otherwise the function is over ridden by other parts of the brain.

2. The floor of the midbrain mainly serves to contain tracts of myelinated nerve fibres connecting the forebrain with the hind brain and spinal cord.

Cerebellum.

1. Receives a sensory input from the eyes, ears, muscles etc.

2. Assesses the conscious desire to move.

3. Controls the co-ordination of muscles needed to carry out desired movement and maintain balance.

All this happens without any conscious awareness. However, if the cerebellum receives conflicting pieces of information a feeling of nausea often results, and balance is difficult to maintain.

Pons

The pons, like the floor of the mid brain, is important for containing myelinated connecting fibres. There are two groups:-

1. Vertical nerve tracts connecting the cerebrum with the medulla.

2. Horizontal nerve tracts connecting the two hemispheres of the cerebellum.

Medulla

This part of the brain is important for controlling processes vital to life, and contains the vital centres, the main ones being:

1. Cardiac centre.

2. Respiratory centre.

3. Vaso-constrictor centre.

Details of how these function are found in the Units on respiration and circulation.

● *Note the vulnerable position of this vital region, just above the hole at the base of the cranium. Breaking the neck will be fatal if it damages the medulla.*

MEMORY

This is a function of the cerebral hemispheres, and it is very much associated with the concept of intelligence. However it is a property of the entire CNS, and it is not possible to denote specific areas to the function of memory. There are two types of memory, referred to as the short and long term memories.

Short term memory.

This is the type of memory that lasts only for a few seconds, and is useful for remembering telephone numbers long enough to dial. To permanently store such information would result in cluttering the brain with too much trivia.

The mechanism is thought to be a re-use of the original neural circuit by which the telephone number, for instance, was read. The circuit loops back on itself allowing the action potentials to loop round and round the same neural path way. The nature of the synapses would tend to result in a gradual drop in frequency, until one synapse fails to reach the required excitory post synaptic potential for firing any action potentials. When the action potentials stop being propagated, the sequence of numbers for that telephone will have been forgotten.

In this theory, it is presumed that there is no permanent change in the synapse. if you want to use the telephone number again you will have to look it up.

Long Term Memory.

This type of memory is permanent. It is possible to put a particular telephone number into the long term memory if it is repeated enough or if the stimulus is strengthened by interest, or importance. This results in the actions potentials looping the circuit more times, eventually bringing about some permanent change in the synapse.

- *Can you think of ways in which the synapse could change?*

Such a transfer into the long term memory takes about half an hour. If an accident makes a person unconscious it is not uncommon to have no recall of the accident, or the time leading up to it.

Once having happened the change is permanent. The telephone number will be permanently remembered. However, if it is not reused it might be difficult finding the right circuit. The telephone number of last year's girl/boy friend might be more difficult to retieve than this year's.

- *As you get close to an examination it is much more use to practise retrieving information than to try pushing more and more in.*

- *The ability to learn a task and remember it can be investiagted by tracing round a star viewed only in a mirror. The effect of viewing a mirror image interfers with the normal hand eye coordination. However this is a task that is quickly learnt. The results can be monitored by measuring time taken, and number of mistakes made.*

SLEEP

Sleep is almost universal among animals, and yet it is not easy to explain or understand. It clearly has a useful survival value in keeping animals safe when darkness would prove unproductive for finding food.

However sleep appears to have a physiological role, and one theory is that chemicals involved in laying down the long term memory are used up within a 24 hour period, and sleep is the time when these chemicals can be resynthesised and accumulated. This idea is supported by the brain activity detected during sleep.

Electroencephalograms (EEG)

These record the electric activity of the brain detected by electrodes placed on the skull. The voltages received are amplified and recorded. The resulting traces show characteristic frequency ranges and rhythms, but not the distinct repetitive patterns produced on the ECG.

Rhythms associated with wakefulness and sleep.

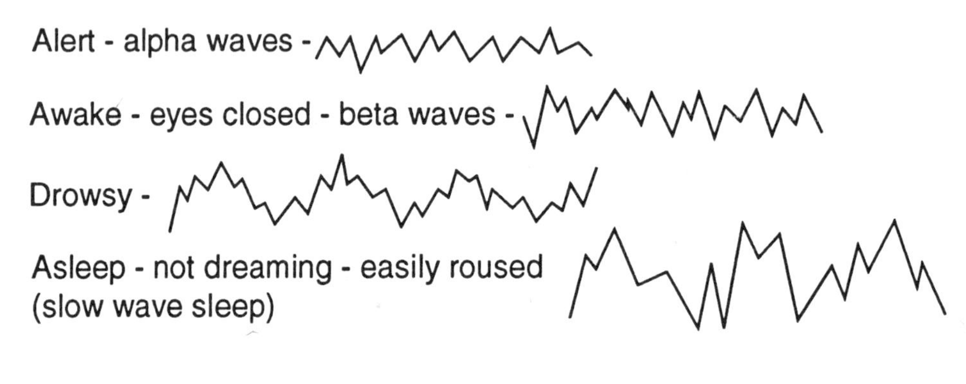

Asleep - dreaming - not easily roused
(rapid eye movement - REM - sleep)

Slow wave sleep and REM sleep alternate during one nights sleep. REM sleep is thought to be important for the resynthesis of brain chemicals.

Reticular Formation

This is a tangled mass of short axoned neurones in the centre of the brain connecting the spinal cord with the higher centres, mainly through the thalamus.

Neurones from the reticular formation, known as the **Reticula Activating System (RAS),** maintain wakefulness. **The Sleep Centres** in the reticula formation have neurones producing the transmitter substance **serotonin.** When serotonin reaches a certain level the RAS system is turned off.

THE PERIPHERAL NERVOUS SYSTEM

This is the part of the nervous system that lies outside the skull and the vertebral column. It connects the receptors of sensory tissues with the CNS, and the CNS with the effecters of the motor tissues. Thus it is in the peripheral system that the long fibres of the sensory and motor neurones are found.

Motor neurones are of 2 types:-

1. **The Somatic Motor Neurones,** which terminate in the striated muscle, and are under voluntary control.
2. **The Autonomic Motor Neurones,** which terminate in the smooth muscles and secretory glands, and are not under voluntary control. These are further subdivided into two types:

a. Sympathetic Motor Neurones.
b. Parasympathetic Motor Neurones.

Most smooth muscles and secretory glands are supplied with both types of autonomic neurone.

The Peripheral Nerves.

These are bundles of fibres that connect with the CNS. The body is divided into somites (or segments), though a degree of specialisation masks this organisation. However, the segmentation is still reflected in the number of peripheral nerves because there is one pair per segment. Thus there are twelve pairs of cranial nerves and 31 pairs of spinal nerves.

- *Conveniently the nerves are known by the number of the segment as well as by a name. On the diagram on page 31 only the numbers are given, unless the name of the nerve is in common use and has been used in other units. Where a number of nerves combine the name of the plexus is given.*

There are three types of nerves:-

1. **Sensory nerves**, where all the fibres in the nerve are sensory.
2. **Motor nerves**, where all the fibres in the nerve are motor.
3. **Mixed nerves**, where both types of fibre are present.

Ganglia

These are bodies where peripheral neurones synapse outside the CNS. They are seen as bulges in the peripheral nerve because of the cell bodies.

DIAGRAM OF THE PERIPHERAL NERVOUS SYSTEM

- *The diagram shows the relationship of the different components that interact with the peripheral system. You might colour the boxes for the sensory and motor components of the peripheral system to link those together.*

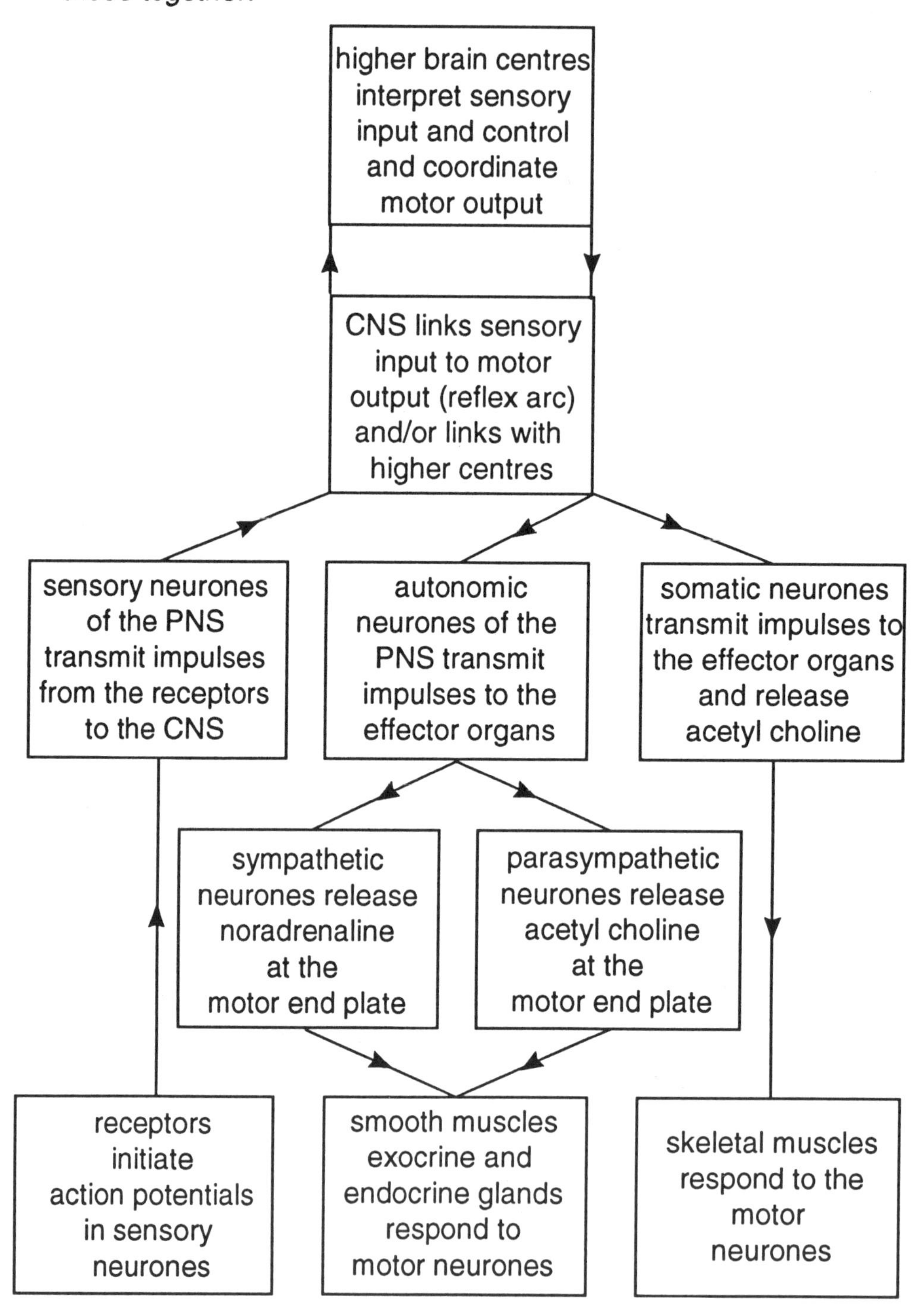

- *The time take to respond to a stimulus is the reaction time. It will depend on the number of neurones involved because of the time taken to cross the synapse. There are always 3 neurones into the CNS and 2 out. The extra neurones will be those of the higher brain centres. Thus thinking adds to the reaction time.*

DIAGRAM OF THE PERIPHERAL NERVES

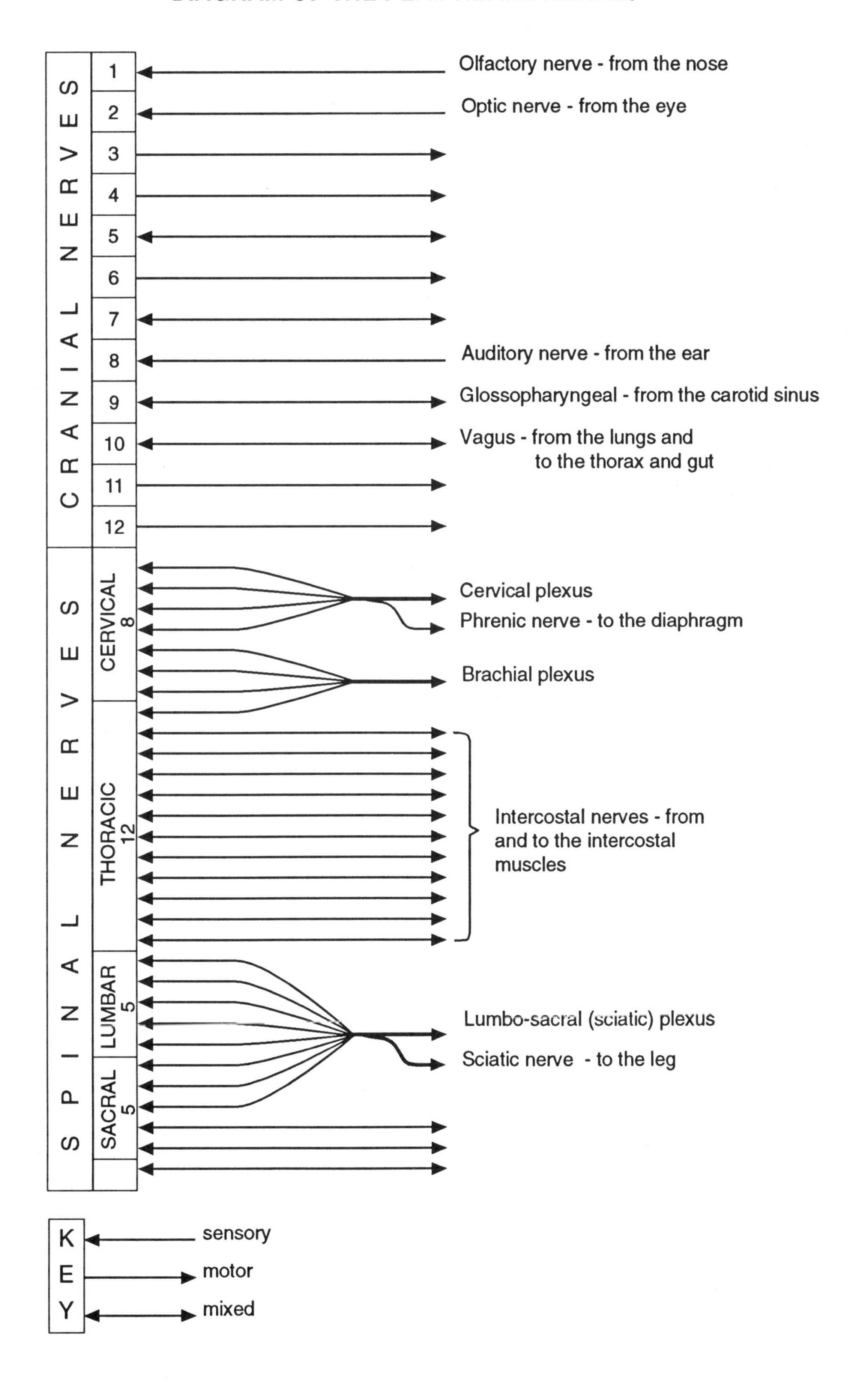

THE SPINAL CORD

This is part of the CNS that is protected by the vertebral column. It is a tube of nervous tissue continuous with the brain, and surrounded by the meninges.

Functions of the spinal cord.

1. To receive the sensory input from the spinal nerves.

2. To send motor output to the spinal nerves.

3. To provide the neural link between the brain and the spinal nerves.

Organisation of the tissues in the spinal cord.

The **grey matter** is the area where the spinal nerves synapse.

The **white matter** is the area where the tracts of myelinated fibres link the brain and the spinal cord.

Ascending fibres are located dorsally and laterally, and synapse with the medulla, cerebellum, or thalamus.

Descending fibres are located ventrally and laterally. The neurones will be the pyramidal cells from the cerebral hemispheres and the neurones from the cerebellum.

- *You could mark in the approximate position of these fibres in the drawing below.*

Transverse Section Across the spinal cord.

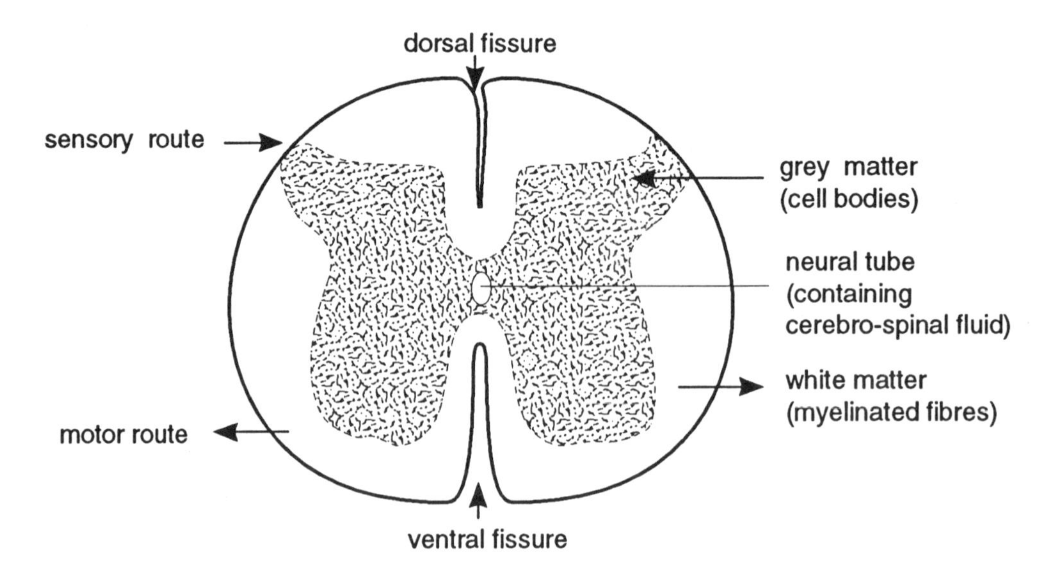

SPINAL REFLEX ACTION

This is a quick automatic response to a stimulus involving only spinal nerves. It is quick because the number of synapses is small. The **knee jerk** in response to a tap on the knee is a well known example.

The sequence of events is:

Stimulus - applied by a sharp tap below the patella

Receptor - in this case the stretch receptors in the muscle with its origin on the femur and its insertion on the tibia. These fire action potentials in the sensory neurone.

Neural Pathway - the impulse travels along the sensory neurone into the spinal cord via the dorsal route where it synapses with the motor neurone. The impulse then travels back to the same muscle along the motor neurone.

Effector - the muscle contracts.

Response - the tibia is pulled up, jerking the leg in the typical knee jerk response.

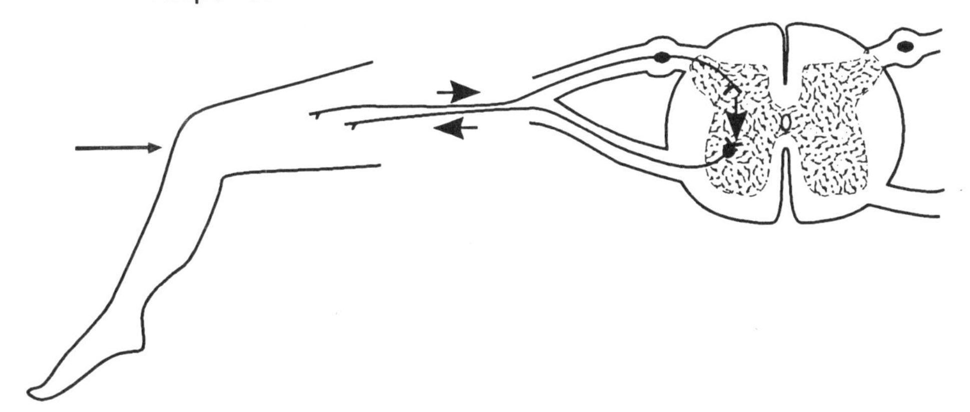

- *Note that the lower leg needs to be swinging free to allow the response to be seen.*

This is a particularly quick response because the sensory neurone synapses directly with the motor neurone. Ascending sensory tracts will also synapse with the sensory neurone, so that we are aware of the stimulus and response. However the speed of the reflex arc makes it a difficult response to inhibit.

Other Spinal Reflexes

Most reflexes have at least one interneurone, such as the response to picking up something hot. The reflex response is to drop the object, even before the heat has been felt. However if the object is a valuable china plate, impulses from the descending motor tracts can result in inhibitory post synaptic potentials which will save the plate.

VOLUNTARY RESPONSE TO A STIMULUS

The cerebral hemispheres will be involved with a voluntary response. Following the neural pathways from the stimulus to the response is complex, but it is possible to analyse the major pathways.

Consider the following scenario:

A man is seated at a desk writing. He hears the door open and arises.

The following summarises the neural consequences of the door opening which leads to the man rising.

1. stimulus - noise of the door opening
2. receptor - generator potential initiated in the sensory cells.
3. action potentials fired in the auditory sensory neurones.
4. action potentials fired in the hind brain.
5. action potentials fired in the thalamus.
6. action potentials fired in the auditory area of the cerebrum
7. action potentials fired in the motor area of the cerebrum.
8. action potentials fired in the spinal cord
9. effector - muscles in the legs contract.
10. response - man stands up

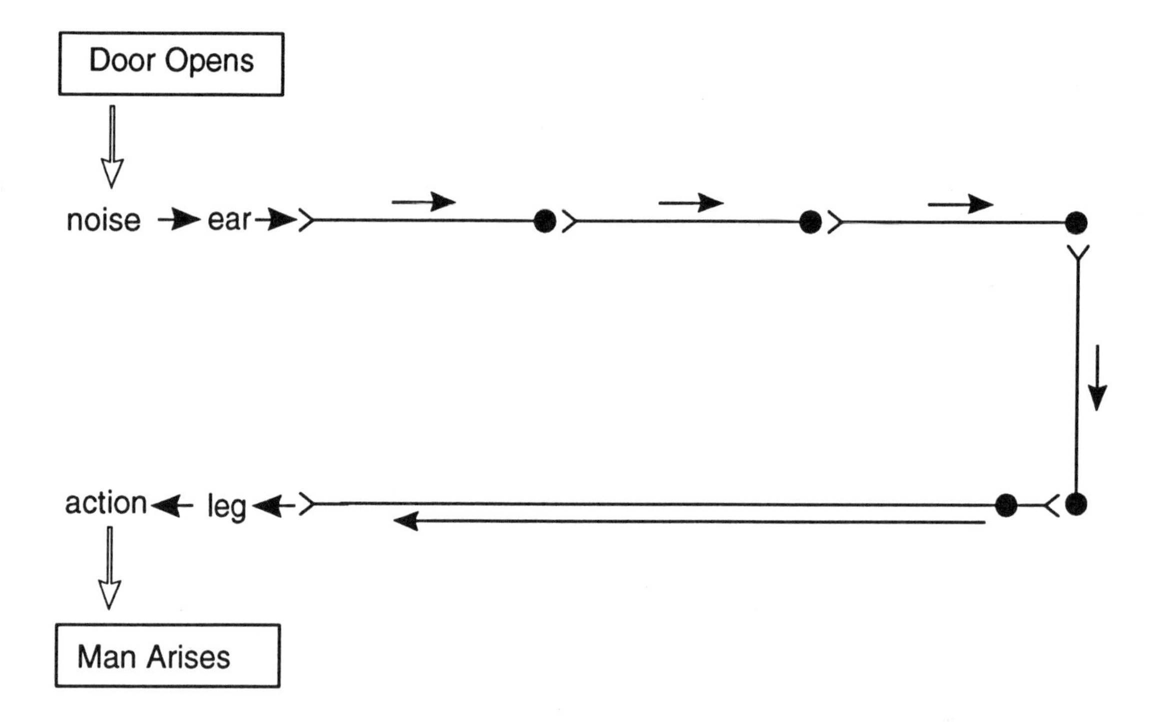

A minimum of 6 neurones would be necessary to complete the circuit, though in practise there would be many many more neurones involved.

● *Use these notes to label the diagram on the next page.*

VOLUNTARY RESPONSE TO A STIMULUS (Cont.)

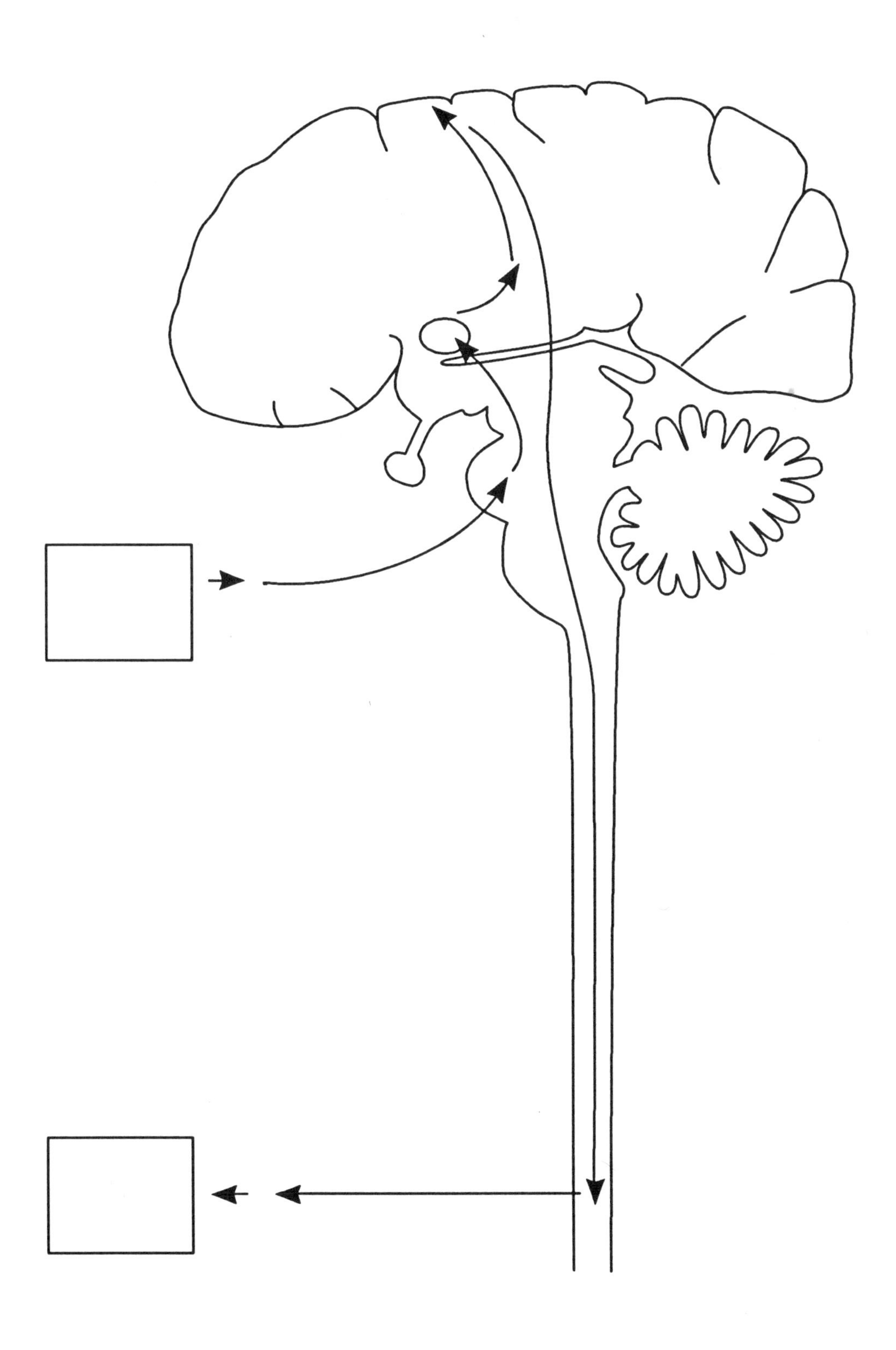

AUTONOMIC NERVOUS SYSTEM

This system automatically controls the effectors that are not under voluntary control.

It has two main functions:

1. To maintain hoeostasis under normal conditions through a negative feedback system. e.g. the homeostatic control of body temperature.
2. To modify the responses to meet the needs of abnormal conditions through a positive feedback system. e.g. the fight or flight response.

The autonomic system is controlled by the hypothalamus.

Mechanism of control by the Autonomic system.

Each effect is supplied with two types of nerve, the sympathetic and the parasympathetic. Each has two neurones between the CNS and the effector. The synapse between these neurones produces a swelling in the nerve called a ganglion. The sympathetic and the parasympathetic nerves are different in terms of the relative length between the pre - and post - ganglionic fibres, the transmitter substance produced at the effector, and the response of the effector.

The diagram below summarises the differences:-

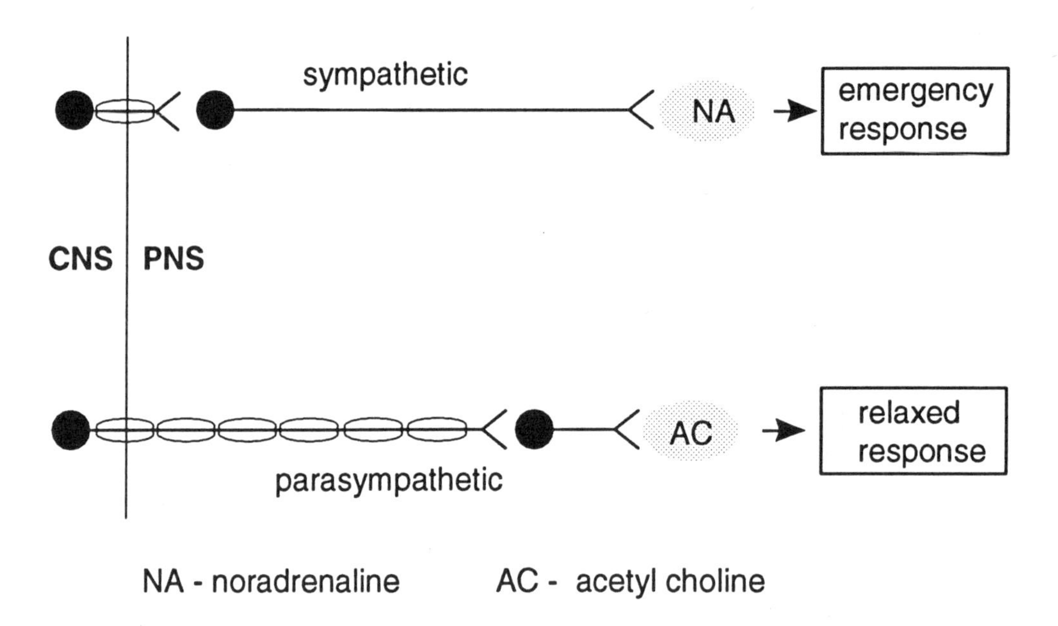

The responses tend to be antagonistic, those of the sympathetic being appropriate to **emergency situations** and to occur simultaneously in all innervated effectors. Those of the parasympathetic are appropriate to **normal relaxed situations** and responses do not necessarily occur simultaneously.

AUTONOMIC NERVOUS SYSTEM (Cont.)

1. Role in Homeostasis

The autonomic system is under the control of the hypothalamus. Both the sympathetic and the parasympathetic nerves will normally carry low frequency action potentials sustaining a normal level of response in the effectors. When the sensory input to the hypothalamus indicates a variation from the norm, then the relative frequencies of the sympathetic and parasympathetic nerves are adjusted by the hypothalamus to restore the norm. This is a negative feedback system.

Temperature regulation is an important negative feedback system controlled by the hypothalamus. The hypothalamus responds to changing blood temperature. A rise in blood temperature results in the hypothalamus turning on the temperature cooling mechanisms. A fall in the blood temperature results in the hypothalamus turning on the mechanisms for reducing heat loss. Most of the mechanisms involved are controlled by the sympathetic neurones to the skin.

2. Role in Abnormal Situations

These could be very frightening situations or very pleasurable situations. The sensory input initiates an increase in frequency of action potentials in the appropriate system and a decrease in the other. If the sensory input continues then these changes continue. This is not a self regulatory system, and it produces changes that deviates from the normal, to meet the demands of the abnormal situation. It is a positive feed back mechanism

e.g. The Fight or Flight Response. Imagine that a lion is about to pounce. Do you stay and grapple, or do you run away? Either way you need all your wits and available strength for survival, and energy consuming processes like digesting the last meal can wait. After all there is no point digesting a meal if you are not going to survive to enjoy the products. The sympathetic nervous system produces a variety of responses, all of which favour survival.

When the maximum level of frequency is reached the sympathetic will be 'turned off' and immediately the parasympathetic will be 'turned full on'. Thus at the height fear there may be pleasurable feeling!

In contrast the parasympathetic activates the same effector organs in a contrasting way, favouring the responses that give pleasure in a relaxing way. Digestion of a meal is favoured by a pleasant relaxing environment.

- *You should consult the text books for specific examples of the responses to the sympathetic and parasympathetic neurones, but a common sense view of what best serves the role of each system will produce the information as well.*

AUTONOMIC REFLEX RESPONSE

It is possible for any part of the autonomic system to bypass the hypothalamus when the neurones from the receptor to the effector form a reflex arc. If the number of synapses involved is small then there will be a quick automatic response to a stimulus involving only one autonomic response.

e.g. the response of the radial and circular smooth muscle of the iris.

Stimulus - dim light
Receptor - rods of the retina

Neural pathway - from the retina via
bipolar cells and
ganglion cells
to the thalamus
from the thalamus
via the sympathetic
to the radial muscle
of the iris

Response - radial muscle contracts

Result - Pupil size increased
to let more light into the eye

Stimulus - bright light
Receptor - cones of the retina

Neural pathway - from the retina via
bipolar cells and
ganglion cells
to the thalamus
from the thalamus
via the parasympathetic
to the circular muscle
of the iris

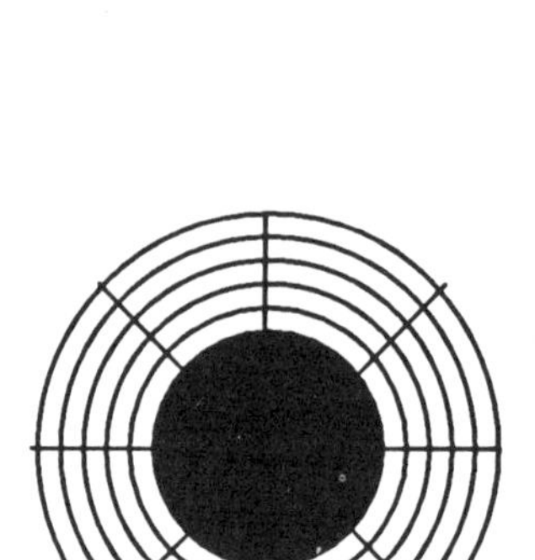

Response - circular muscle contracts

Result - Pupil size decreased
to reduce the light entering
the eye

Conditioned Autonomic Reflex

Any autonomic reflex can be conditioned to occur in response to different stimuli by a learning process. The classical experiments in this area were performed by the Russion, Pavlov. The autonomic reflex he investigated was the production of saliva which normally increases in the presence of food. He used dogs in his experiments, and by presenting a variety of stimuli with the food the dogs learnt to produce saliva in response to, for instance, a bell, even when the food was not present.

THE NERVOUS SYSTEM ASSIGNMENT

Fill in the answers for Section 1 in the spaces provided. Section 2 and 3 should be answered on separate sheets. The assignment is worth 100 marks.

Section 1

1. The diagram shows a section of the spinal cord.

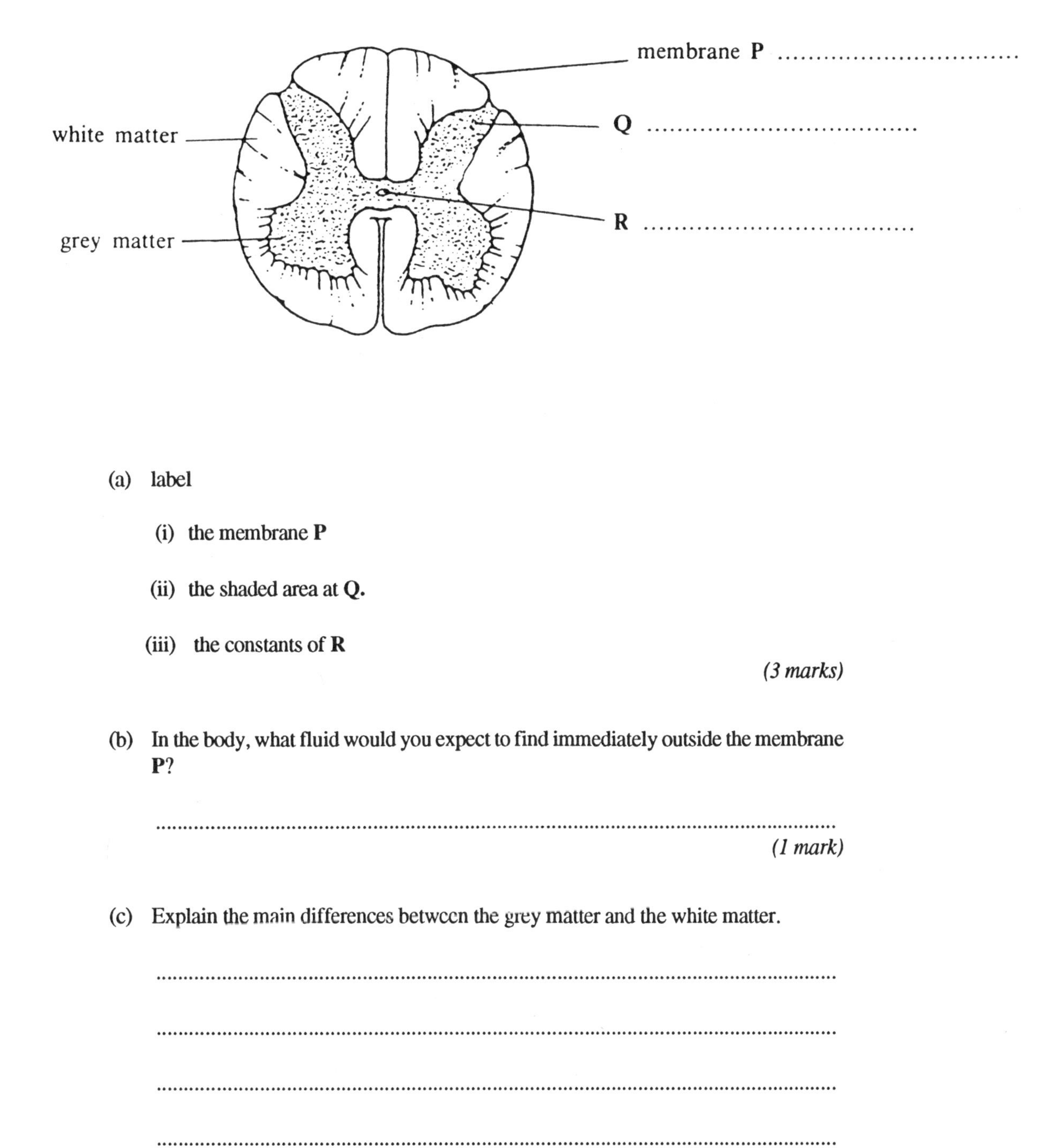

(a) label

(i) the membrane **P**

(ii) the shaded area at **Q.**

(iii) the constants of **R**

(3 marks)

(b) In the body, what fluid would you expect to find immediately outside the membrane **P**?

....

(1 mark)

(c) Explain the main differences between the grey matter and the white matter.

....

....

....

....

(4 marks)

Guidance Note 1: Refer to page 32.

2. The diagram shows a typical human neurone.

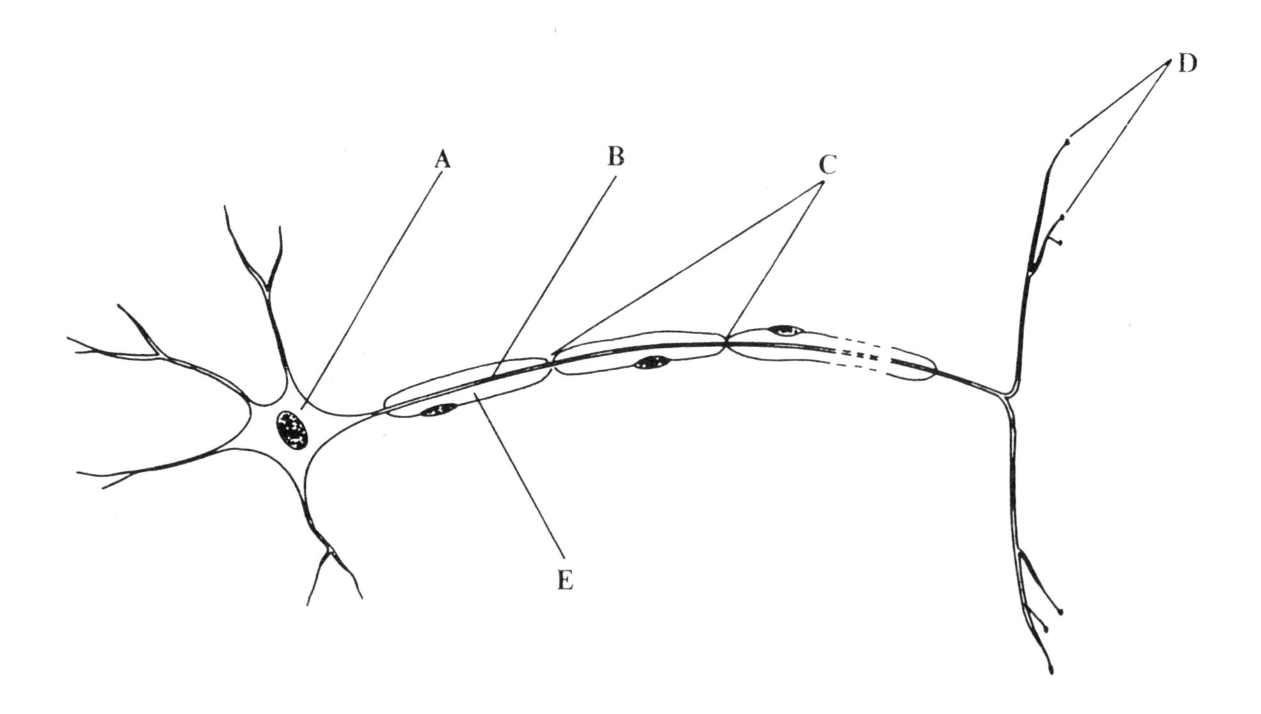

(a) In the spaces below, name the features labelled **A** to **D.**

A ....

B ....

C ....

D ....

(4 marks)

(b) Name the lipid material found in feature **E**

....

(1 mark)

(c) Describe the part played by feature **E** in impulse propagation.

....

....

....

(2 marks)

(d) The diagram shows the arrangement of motor neurones in part of the autonomic nervous system:

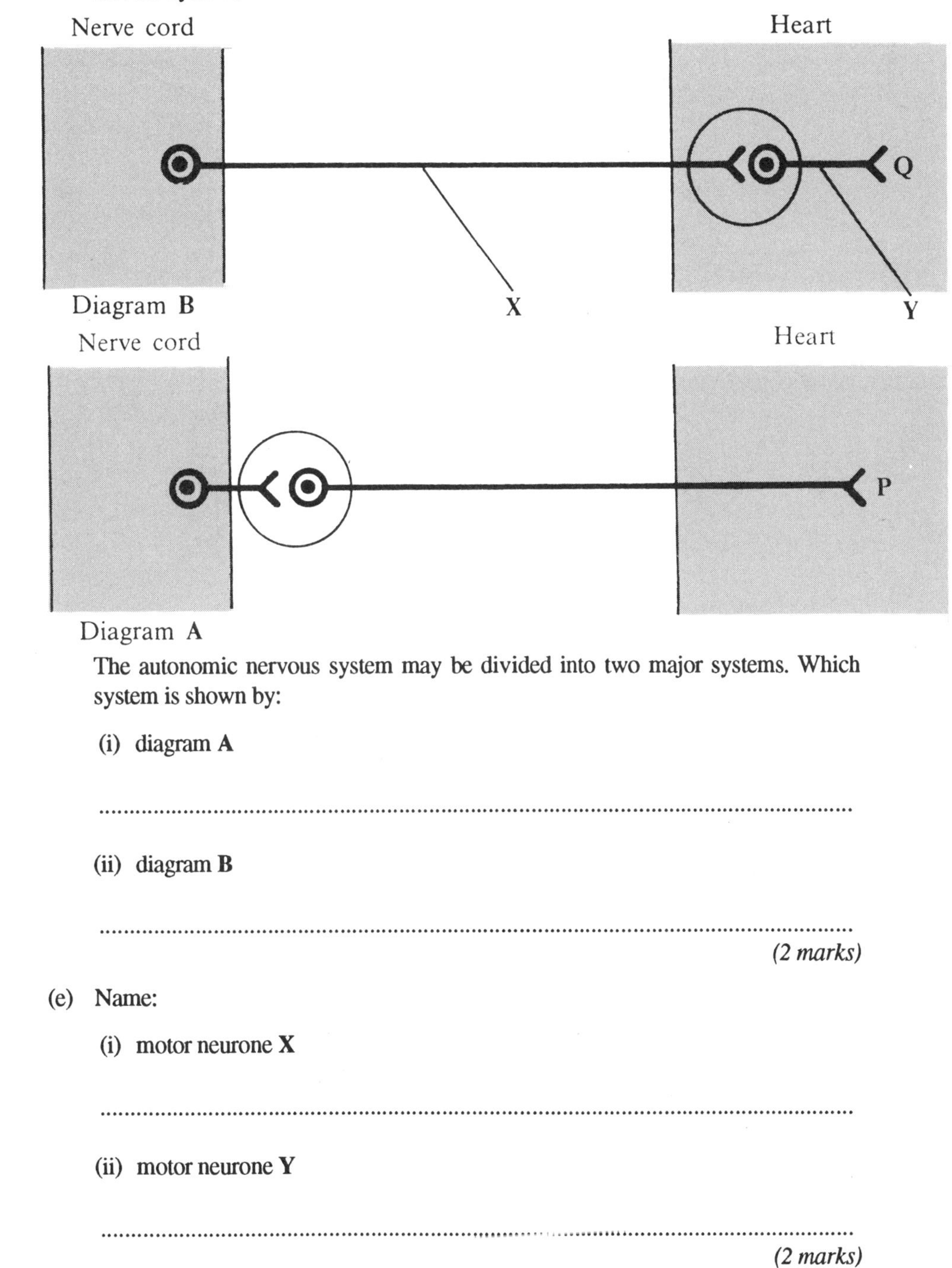

Diagram **B**

Diagram **A**

The autonomic nervous system may be divided into two major systems. Which system is shown by:

(i) diagram **A**

....

(ii) diagram **B**

....

(2 marks)

(e) Name:

(i) motor neurone **X**

....

(ii) motor neurone **Y**

....

(2 marks)

(f) Name the neuro-transmitters released at axon ending:

(i) **P**..

(ii) **Q**..

(2 marks)

(AEB 1988)

Guidance Note 2: Refer to pages 5,6,11, and 36-38.

3. *(6 marks)*

The diagram shows a longitudinal section through the human brain.

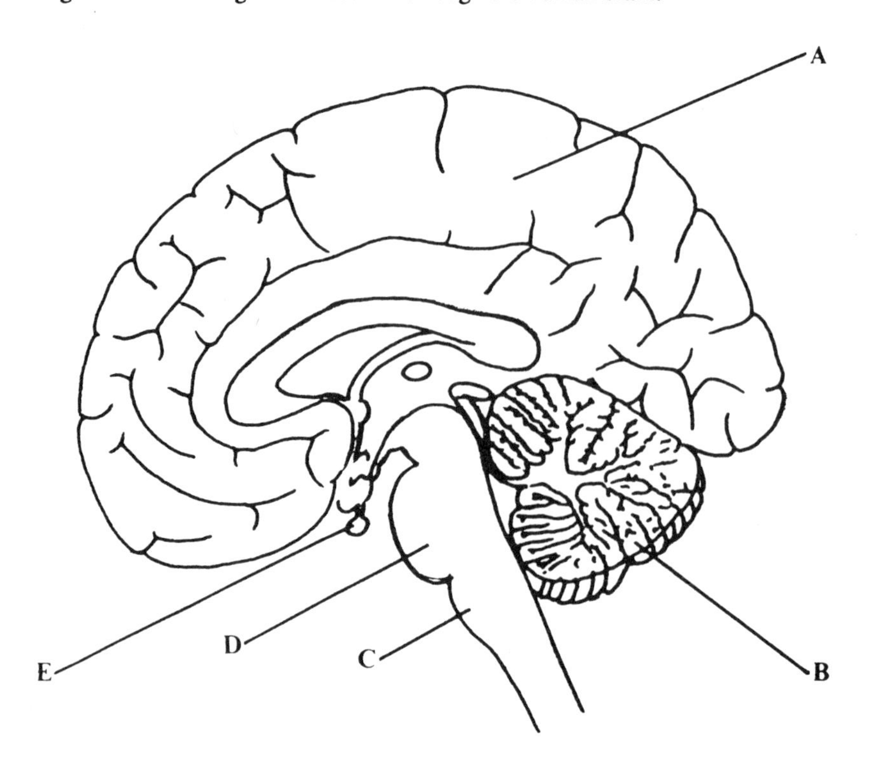

(a) Identify structures **A** and **D**.

A..

B..

(2 marks)

(b) Give **one** function for each of the structures **C** and **E.**

C..

D..

(2 marks)

(c) Suggest **one** functional relationship between structures **A** and **B**

....

....

....

(2 marks)

(AEB 1989)

Guidance Note 3:

4. *(2 marks)*

(a) Describe one location where you would find the **choroid plexus.**

....

(b) What is the function of the choroid plexus?

....

(2 marks)

5. *(6 marks)*

(a) Briefly explain these statements about the nervous conduction.

(i) In a neurone pathway there is a delay of approximately 0,5 milliseconds in transmission time at a synapse.

....

....

....

(2 marks)

(ii) An impulse can pass along an axon in either direction but it can be transmitted across a synapse in only one direction.

....

....

....

(2 marks)

(iii) Myelinated neurones conduct impulses faster than non-myelinated neurones.

....

....

(1 mark)

(b) Suggest **one** effect on skeletal muscle of a drug that inhibits the enzyme acetyl-cholinesterase.

....

....

(1 mark)

(AEB 1991)

Guidance Note 5: Acetyl-cholinesterase is the enzyme that breaks down acetyl choline. You will need to hunt through pages 9-15 for answers.

6. *(7 marks)*

Complete the table to show the effects of the sympathetic and parasympathetic nervous systems. In some instances there may be no effect.

Parts innervated	Sympathetic action	Parasympathetic action
Circular muscles of iris		
Anal sphincter muscle		
Bladder		
Bronchial muscle		
Intestinal muscle		
Hair muscle (erector pili)		
Sweat glands		

7. *(9 marks)*

The following is an account of an investigation into certain aspects of human behaviour.

A strip of metal foil was shaped into a flat, five-pointed star and stuck to a wooden board. the person taking part in the investigation could see the star only by looking into a mirror and was asked to run a metal pointer round the star. Each time contact was lost, a pen recorder marked a vertical line on a moving chart. The records of ten consecutive trials, with a ten second rest after each trial, are shown. For each trial, three records show (i) duration in seconds (ii) number of errors (iii) times when each error occurs.

Records from moving chart of 10 consecutive trials.

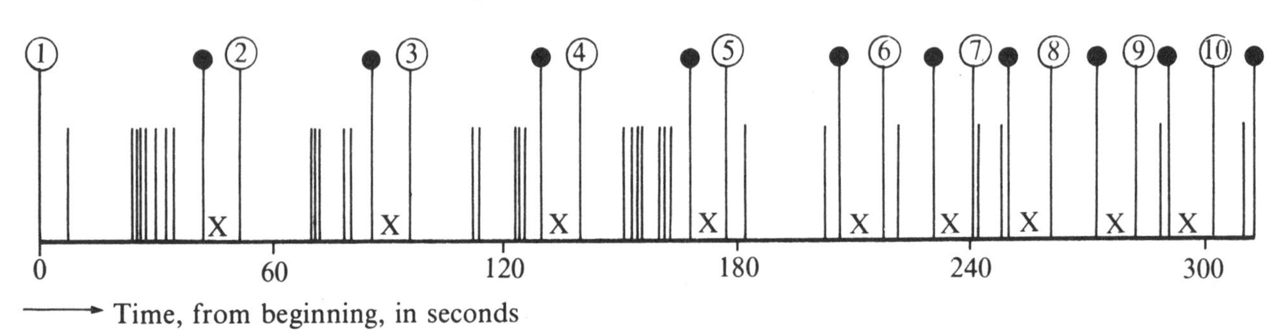

(a) Summarise the change in performance revealed by the data.

....

....

(2 marks)

(b) Predict the final performance (duration and errors) had the person continued for another ten trials. Give reasons for your answers.

....

....

....

....

(4 marks)

(c) Suggest a method for investigating the performance of the person's improvement in ability to carry out this task of tracing a five-pointed star.

....

....

(1 mark)

(d) The skill achieved might be confined to precisely this task or might apply to a range of this general type. State simple modification of the procedure described so you could attempt to discover more about this.

....

(2 marks)

(AEB 1985)

Guidance Note 7: Spend time on interpreting the records before answering the questions.

Section 2

8. Figure **2.1** represents a simple spinal reflex pathway.

Figure **2.1**

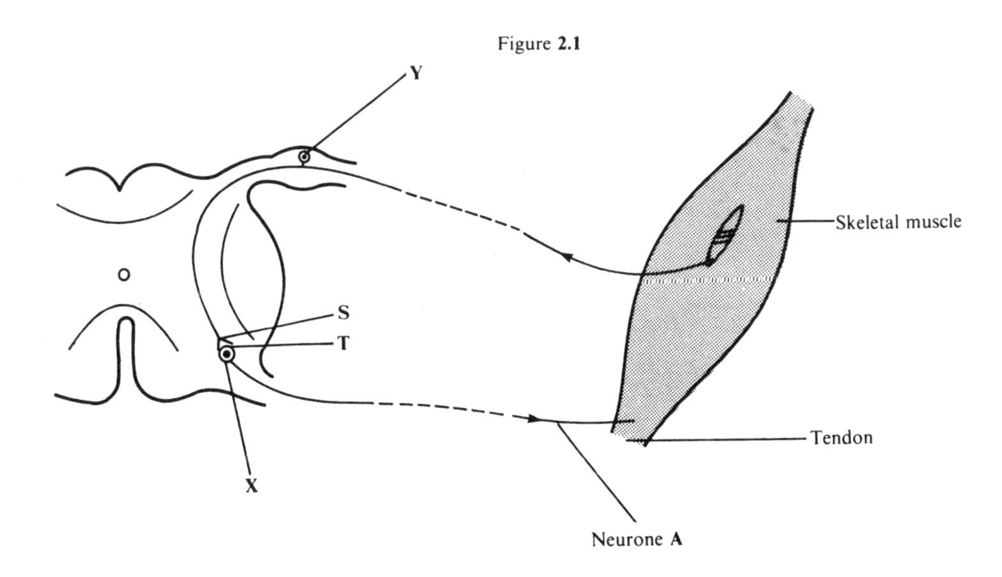

(a) (i) Identify feature **X** and feature **Y** in the diagram *(2 marks)*
(ii) Name the type of sensory receptor in this reflex pathway. *(1 mark)*
(iii) Explain how an impulse is transmitted from point **S** to point **T** in this pathway *(4 marks)*

(b) Distinguish between a simple reflex and a conditioned reflex. *(1 mark)*

(c) An experiment was performed to measure the speed of conduction of impulse in a myelinated neurone supplying a skeletal muscle. The nerve was stimulated at varying distances from the muscle. At each distance, the time delay between the application of a stimulus and the contraction of the muscle was recorded. the results are shown in figure **2.2.**

Figure **2.2**

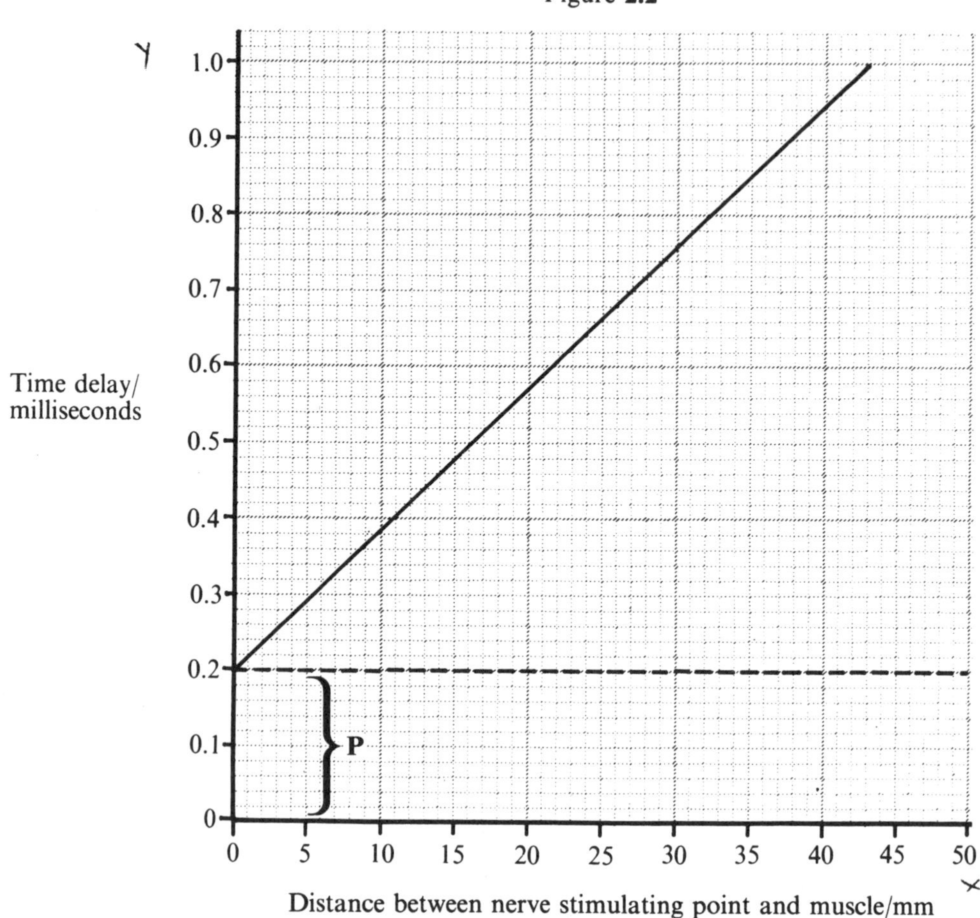

(i) Calculate the speed of conduction of the neurone. Show your working. *(2 marks)*

(ii) Give **three** factors that contribute to the time delay between the application of the stimulus and the contraction of the muscle. *(3 marks)*

(iii) Why is the time delay **P** constant regardless of the distance between the stimulating point and muscle? *(1 mark)*

(d) How would the value obtained for speed of conduction be affected, if at all, if the neurone were:

(i) unmyelinated but the same diameter;
(ii) myelinated but half the diameter?

(2 marks)

(e) Figure **2.3** shows a recording of a muscle fibre action potential when the skeletal muscle was stimulated via its nerve.

Figure **2.3**

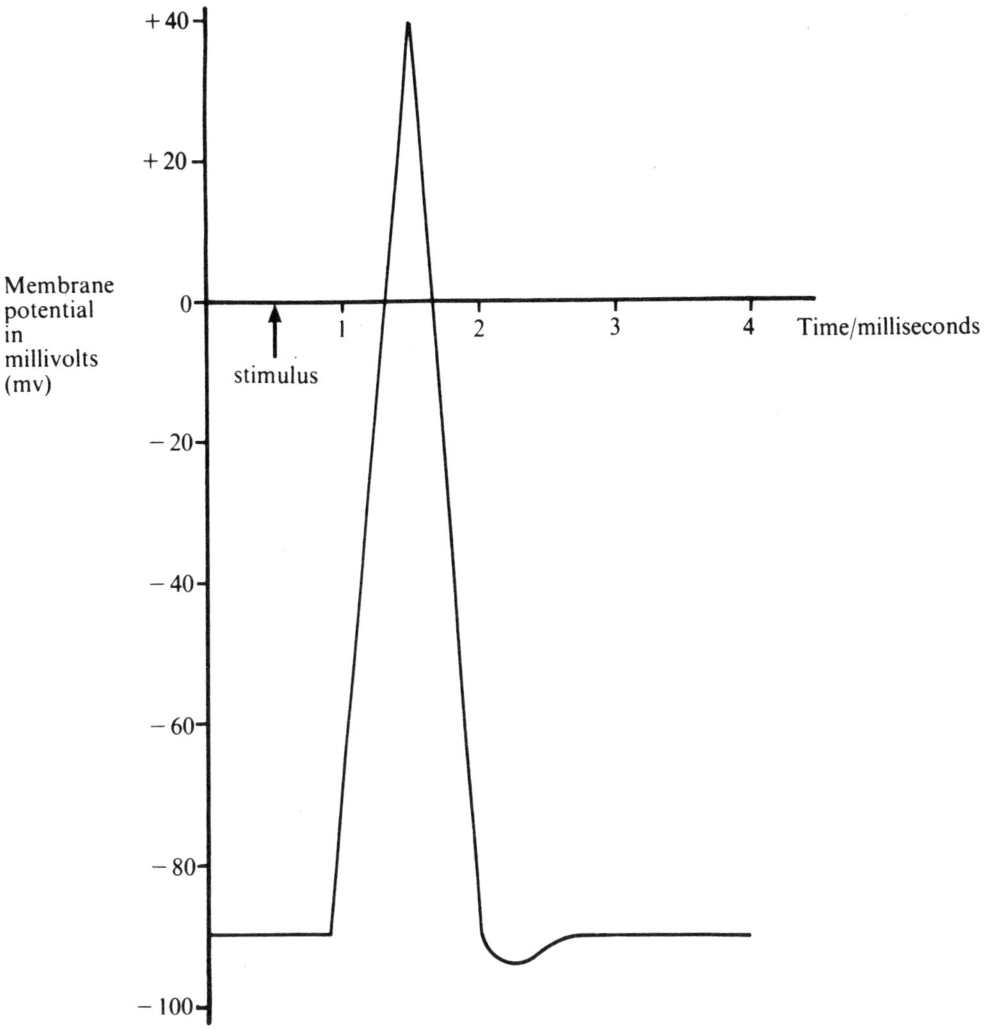

+40
+20
Membrane potential in millivolts (mv)
0
1 2 3 4 Time/milliseconds
stimulus
−20
−40
−60
−80
−100

(i) The recording shows a negative membrane potential of -90mV when the muscle cell is at rest (unstimulated). Explain the cause of the resting membrane potential. *(3 marks)*

(ii) Explain the change in membrane potential from -90mV to +40mV following the stimulation. *(1 mark)*

(f) (i) What is meant by the *absolute refractory period* of a muscle? *(1 mark)*

(ii) Unlike skeletal muscle, which has an absolute refractory period of 1.5 milliseconds, cardiac muscle has an absolute refractory period of 200 milliseconds. Explain the advantage of this to the heart. *(1 mark)*

(AEB 1992)

Guidance Note 8: This question is testing comprehension, you should try thinking out the answers, not looking them up. Note that 'show your working' means explain what you are doing.

Section 3

9. Give an account of the ways in which the peripheral nervous system is adapted to its function.

(25 marks)

(AEB 1987)

Guidance Note 9: You should remeber to include the sensory input, as well as motor output, and the cranial as well as spinal nerves. The more sophisticated answers would be organised into the **ways** it is adapted. Try jotting ideas down before you begin.